T0351313

Principles of Synthetic Aperture Radar Imaging

A System Simulation Approach

Signal and Image Processing of Earth Observations Series

Series Editor

C.H. Chen

Published Titles

Principles of Synthetic Aperture Radar Imaging: A System Simulation Approach
Kun-Shan Chen

Remote Sensing Image Fusion
Luciano Alparone, Bruno Aiazzi, Stefano Baronti, and Andrea Garzelli

Principles of Synthetic Aperture Radar Imaging

A System Simulation Approach

Kun-Shan Chen

CRC Press
Taylor & Francis Group
Boca Raton London New York

CRC Press is an imprint of the
Taylor & Francis Group, an **informa** business

CRC Press
Taylor & Francis Group
6000 Broken Sound Parkway NW, Suite 300
Boca Raton, FL 33487-2742

© 2016 by Taylor & Francis Group, LLC
CRC Press is an imprint of Taylor & Francis Group, an Informa business

No claim to original U.S. Government works

Printed on acid-free paper
Version Date: 20151022

International Standard Book Number-13: 978-1-4665-9314-5 (Hardback)

Visit the Taylor & Francis Web site at
http://www.taylorandfrancis.com

and the CRC Press Web site at
http://www.crcpress.com

In loving memory of my mother

Contents

Preface

The purpose of this book is to provide a systematic explanation of modern synthetic aperture radar (SAR) principles for the use, study, and development of SAR systems. The framework and scope of this book are suitable for students, radar engineers, and microwave remote-sensing researchers.

SAR is a complex imaging system that has many sensing applications, from geoscience studies to planetary exploration. In the last two decades, many excellent books in treatment of one or more specific topics on SAR have been published.

This book is unique not in presenting novel topics but in its inclusion of chapters on signal speckle, radar-signal models, sensor-trajectory models, SAR-image focusing, platform-motion compensation, and microwave-scattering from random media. Numerical simulations are presented to cover these important subjects. A ground-based FMCW system is also included. And, as an example, system simulation is provided for target classification and recognition. Simulation flowcharts and results are graphically presented throughout this book to help the reader, if necessary, to grasp specific subjects. The following paragraphs of this preface will elaborate on the aforementioned subjects.

Many publications have sought to connect the wave-scattering process and the radar-signal process by presenting physical, systemic, and signal SAR models, among others. In this book, numerical simulations of polarimetric wave scattering from randomly irregular surfaces are given to illustrate the speckle phenomenon and to validate the fully developed speckle model for a homogeneous target. However, it is logical to extend numerical simulations to volume and surface–volume scattering for more complex targets. The signal model using a point-spread function is given for chirp and FMCW system. This kind of model greatly simplifies the data process but profoundly neglects the coherence within the resolution cell, where many targets are in presence. It remains challenging to work on a full-blown signal model to account for the target homogeneity and the memory effects. This topic will be reserved for future research. For demonstration, we have selected range-Doppler and chirp-scaling algorithms because of their popularity and their high precision. In path trajectory, transformations of time and space coordinates of aircrafts and satellites are presented, the latter being more complicated because of the effect of earth's rotation. For SAR systems, time and space coordinates are critical. Doppler frequency and Doppler-frequency rates are two other vital parameters for precise and coherent processing. A simple aircraft trajectory with bias and noise is simulated in this book. This proves useful for accounting for platform motion and image focusing, subjects that are treated in Chapters 5 and 6. Chapter 7 specifically deals with a ground-based SAR. Because of the short range of the operation, FMCW is adopted. Suitable focusing algorithms using range-Doppler and chirp scaling are presented. An experimental system operating at Ka-band was built in a microwave aniconic chamber to demonstrate data collection and image focusing. Chapter 8 includes topics from previous chapters to illustrate the complete chain of a SAR operation. Such system simulations are extremely useful in understanding image properties and in developing user applications.

Acknowledgments

Many pioneers in SAR research and development inspired me to explore this fascinating field. Their names and works will appear in references throughout this book. In many ways, it is not possible to include them all. However, I would like to acknowledge here a few of those people who have been important to my work. In my last two decades of teaching and research on microwave remote sensing, I have greatly benefited from the publications of Professor A. K. Fung, R. K. Moore, Professor F. T. Ulaby, Professor J. A. Kong, and Professor L. Tsang. Professor W. M. Boerner offered great advice and inspiration in many aspects of SAR, in particular, POLSAR. Dr. Jong-Sen Lee, as a mentor and friend, guided me through the basics of SAR, including speckle modeling and their filtering and polarimetric SAR and its applications. I am enormously grateful for our rewarding discussions during his academic visits. Professor H. D. Guo, the director general of RADI, CAS, has been a valuable source of consultation in SAR geoscience applications. I greatly appreciate his generous help.

The material presented in this book is mostly selected from my lecture notes and research project reports. Many of my former students participated in these studies, in one way or another. The computer simulations were mostly implemented and produced by Steve Chiang, whom I deeply appreciate. Dr. Shuce Chu provided excellent work on experimental measurements, which took tremendous efforts. Dr. Chih-Tien Wang greatly contributed in image focusing and processing. My research secretary, Peiling Chen, provided both administrative and technical support. I owe them all my sincere thanks.

I must thank my book series editor, Professor C. H. Chen, who encouraged me to write this book. His input and advice were invaluable for this project. I would also like to extend my gratitude to my editor, Irma Britton, and her team members at CRC Press. Her patience and assistance have made this book possible. Last, but not least, I am indebted to my wife, Jolan, for her constant support and encouragement, and to my three children, Annette, Vincent, and Lorenz, for their understanding of my constant absence from home.

Author

Kun-Shan Chen received a PhD degree in electrical engineering from the University of Texas at Arlington in 1990. From 1992 to 2014, he was with the faculty of National Central University, Taiwan, where he held a distinguished chair professorship from 2008 to 2014. He joined the Institute of Remote Sensing and Digital Earth, Chinese Academy of Science, in 2014, where his research interests include microwave remote sensing theory, modeling, system, and measurement from terrain with applications to environmental watch. He is also associated with the Department of Electrical Engineering, The University of Texas at Arlington, as a research professor since 2014. He has authored or coauthored more than 110 referred journal papers, has contributed 7 book chapters, and is a coauthor (with A. K. Fung) of *Microwave Scattering and Emission Models for Users* (Artech House, 2010). His academic activities include being a guest editor for the IEEE TGARS Special Issue on Remote Sensing for Major Disaster Prevention, Monitoring and Assessment (2007), a guest editor for the Proceedings of IEEE Special Issue on Remote Sensing for Natural Disaster (2012), an IEEE GRSS ADCOM member (2010–2014), a founding chair of the GRSS Taipei Chapter, an associate editor of the IEEE Transactions on Geoscience and Remote Sensing since 2000, and a founding deputy editor-in-chief of the IEEE Journal of Selected Topics in Applied Earth Observations and Remote Sensing (2008–2010). A fellow of IEEE (2007), Dr. Chen served since 2014 as a member of the editorial board of the Proceedings of the IEEE.

1 Preliminary Background

Since its early development in the 1950s, the synthetic aperture radar (SAR) has evolved into a powerful and valuable tool for monitoring our earth environment [1–16]. SAR is a complex device that requires deliberate integration of sensors and platforms into a fully unified system so that the imagery data it acquires are sufficiently accurate, both radiometrically and geometrically. The rapid progress of hardware and software has driven the SAR system to produce superb quality images like never before. The abundance of repeat-pass interferometric and fully polarimetric data has opened new domains of application since 1990, making SAR an indispensable instrument for diagnosing and monitoring natural and man-made disasters [2,7].

This book focuses on SAR imaging from a systems point of view, albeit one of signal aspects, with a complete chain from wave–target interactions to image formation to target feature enhancement. SAR is a complex system that measures the scattered wave of a target under an impinging incident wave that is transmitted by setting the probing frequency, polarization, and observing geometry. Essentially, SAR involves radar operation and image formation. The key to SAR is to obtain the Doppler shifts embedded in the carrier frequency, by moving either the radar or the target being observed. Thus, through proper operation and data processing, SAR can accurately obtain a target's reflectivity map and velocity (if not stationary).

1.1 SIGNALS AND LINEAR SYSTEM

Arbitrary signal $x(t)$ can be broken down by [17]

$$x(t) = \int_{-\infty}^{\infty} x(\tau)\delta(t-\tau)\,d\tau \tag{1.1}$$

If $h(t)$ is the response at time t due to input of $\delta(\tau)$, then the output from the signal $x(t)$ is

$$y(t) = \int_{-\infty}^{\infty} x(\tau)h(t,\tau)\,d\tau \tag{1.2}$$

$h(t)$ is called the impulse response. For practical cases, such as the radar system discussed in this book, the system is causal, that is, $h(t, \tau) = 0$, $t < \tau$, $t < \tau < \infty$. Also, if $h(t, \tau) = h(t - \tau)$,

$$y(t) = \int_{-\infty}^{\infty} x(\tau)h(t-\tau)\,d\tau = \int_{-\infty}^{\infty} h(\tau)x(t-\tau)\,d\tau \tag{1.3}$$

Equation 1.3 indicates the system is time invariant. Also, we assume the system is linear (Figure 1.1). Any signal may be reproduced by

$$\int_{-\infty}^{\infty} x(t)\delta(t-t_0) = x(t_0) \tag{1.4}$$

$$\int_{a}^{b} x(t)\delta(t-t_0)\,dt = x(t_0),\ a \le t_0 \le b \tag{1.5}$$

Equation 1.4 or 1.5 is actually a signal with sampling function $\delta(t)$, a delta function with

$$\int_{-\infty}^{\infty} \delta(t)\,dt = 1 \tag{1.6}$$

If the impulse response function is a delta function, the signal can be perfectly reconstructed—an ideal situation. Expansion to a multidimensional signal is readily straightforward. There are alternative expressions of the delta function that are useful in other cases [17,18].

If

$$\int_{-\infty}^{\infty} \left|x(t)\right|^2 dt < \infty \tag{1.7}$$

then there exists its Fourier transform:

$$X(\omega) = \int_{-\infty}^{\infty} x(t)e^{-j\omega t}\,dt \tag{1.8}$$

where $X(\omega)$ is a complex spectrum with

$$X(\omega) = X_r(\omega) + jX_i(\omega) = A(\omega)e^{j\theta(\omega)} \tag{1.9}$$

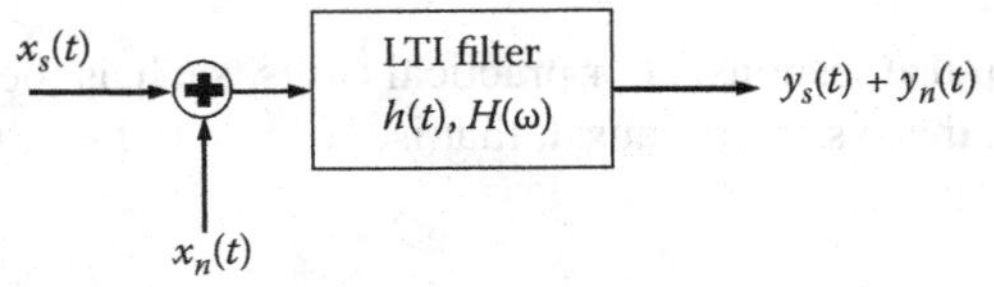

FIGURE 1.1   Linear time-invariant (LTI) system.

For a band-limited signal, such as the radar signal we deal with,

$$X(\omega) = 0, |\omega| > B \tag{1.10}$$

$$x(t) = \frac{1}{2\pi} \int_{-B}^{B} X(\omega)e^{j\omega t}\, d\omega \tag{1.11}$$

$$x(t) = 0, |t| < T \tag{1.12}$$

$$X(\omega) = \int_{-T}^{T} x(t)e^{-j\omega t}\, dt \tag{1.13}$$

The appendix gives special transform pairs and relationships that are useful in later signal analyses throughout this book.

1.1.1 ANALYTIC SIGNAL

In signal analysis, a real-valued signal can be represented in terms of the so-called analytic signal from which the amplitude, phase, and frequency modulations of the original real-valued signal can be determined [17,19,20]. If a real-valued $x(t)$ is

$$x(t) = \frac{1}{2\pi} \int_{-\infty}^{\infty} X(\omega)e^{j\omega t}\, d\omega \tag{1.14}$$

and if $s(t)$ denotes the analytic signal of $x(t)$, then

$$s(t) = \frac{1}{\pi} \int_{0}^{\infty} X(\omega)e^{j\omega t}\, d\omega \tag{1.15}$$

It is possible to obtain $s(t)$ in terms of $x(t)$, in which $s(t)$ is the so-called analytic signal:

$$S(\omega) = 2U(\omega)X(\omega) \tag{1.16}$$

where $U(\omega)$ is the Heaviside step function (see Equation A.7).

$$u(t) = \mathcal{F}^{-1}[U(\omega)] = \frac{1}{2}\delta(t) + j\frac{1}{2\pi t} \tag{1.17}$$

From the convolution property (Equation A.7) we have

$$s(t) = 2u(t) \otimes x(t) = x(t) \otimes \left[\delta(t) + \frac{j}{\pi t} \right] = x(t) + jq(t) \qquad (1.18)$$

where from Equation A.11, we have $q(t) = \dfrac{1}{\pi t} \otimes x(t)$.

The analytic signal associated with a real-valued function f can be obtained by its Hilbert transform to provide the quadrature component [19,21].

Assuming $x(t)$ is Fourier transformable, $x(t) \Leftrightarrow X(\omega)$; thus, Equation 1.3 can be written as

$$Y(\omega) = \int_{-\infty}^{\infty} y(t)e^{-j\omega t}\, dt = X(\omega)H(\omega) \qquad (1.19)$$

where $h(t) \Leftrightarrow H(\omega)$. It is noted that for $x(t) \in \mathbb{R}$, $X(-\omega) = X^*(\omega)$.

In general, Parseval's theorem states that

$$\int_{-\infty}^{\infty} |x(t)|^2\, dt = \frac{1}{2\pi} \int_{-\infty}^{\infty} |X(\omega)|^2\, d\omega \qquad (1.20)$$

From Shannon's sampling theorem, any Fourier transformable signal $x(t)$, all of whose spectral components are at frequencies less than B, can be completely determined in terms of its values at discrete time spacing $\dfrac{1}{2B}$:

$$x(t) = \sum_{n=-\infty}^{\infty} x\left(\frac{n}{2B} \right) \frac{\sin(2\pi Bt - n\pi)}{2\pi Bt - n\pi} \qquad (1.21)$$

The sampling theorem (Equation 1.21) is of profound importance because it allows one to represent the continuous band-limited signal by a discrete sequence of its sample without loss of information.

A radar signal is embedded in a noisy environment [1,18,22]; the echo is a coherent sum of the desired signal and noise. One major concern is how to design a receiving filter such that noise effects are minimized at output. Consider Figure 1.1; we want to determine the filter impulse response function $h(t)$ or the transfer function $H(\omega)$.

The outputs of the signal and noise at the filter are, respectively,

$$y_s(t) = \int_{-\infty}^{\infty} h(\tau) X_s(t-\tau) d\tau \qquad (1.22)$$

$$y_{n(t)} = \int_{-\infty}^{\infty} h(\tau) . X_n(t-\tau) d\tau \qquad (1.23)$$

Assuming $x_n(t)$ is a stationary random noise with power spectrum $S_{x_n}(\omega)$, $y_n(t)$ will also be a stationary random noise with power spectrum [23]

$$S_{y_n}(\omega) = |H(\omega)|^2 S_{x_n}(\omega) \qquad (1.24)$$

The criterion used to determine $h(t)$ or $H(\omega)$ is to maximize the signal-to-noise ratio:

$$H(\omega) = CX_s^*(\omega)e^{-j\omega t_m} \qquad (1.25)$$

$$h(t) = Cx_s(t_m - t) \qquad (1.26)$$

The response with Equation 1.25 or 1.26 is called the matched filter. To be physically realizable, a necessary and sufficient condition for $H(\omega)$ is the Paley–Wiener criterion [20]:

$$\int_{-\infty}^{\infty} \frac{|\ln(|H(\omega)|)|}{1+\omega^2} d\omega < \infty \qquad (1.27)$$

The output of the ideal matched filter is

$$y_{MF}(t) = C \int_{-\infty}^{\infty} x_s(t_m - \tau)x_s(t-\tau) d\tau \qquad (1.28)$$

In Equation 1.28, t_m is the time where $x(t)$ has its peak value.

1.2 BASICS OF RADAR SIGNAL

1.2.1 SINGLE PULSE

A typical radar signal is of the form [10,22]

$$s(t) = m(t) \cos(\omega_c t + \phi) \qquad (1.29)$$

where $\omega_c = 2\pi f_c$, $m(t)$ is the modulating (pulse) waveform, f_c is the carrier frequency, and ϕ is the phase angle.

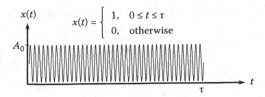

FIGURE 1.2 Single pulse with no modulation.

An analytic form is

$$s(t) = m(t)e^{j[\omega_c t + \phi(t)]}$$ (1.30)

The waveform may determine the range resolution, Doppler resolution, signal-to-noise ratio, range–Doppler coupling, and others (Figure 1.2). A typical modulating waveform is a pulse:

$$m_0(t) = A_0[u(t - \tau) - u(t - \delta - \tau)]$$

$$M_0(\omega) = \mathcal{F}[m_0(t)] = A_0 e^{-j\omega\left(\tau + \frac{\tau}{2}\right)} \frac{2\sin\omega\frac{\tau}{2}}{\omega}$$ (1.31)

The Fourier integral of the output signal is

$$S_0(\omega) = A_0 e^{-j(\omega - \omega_c)\left(\delta + \frac{\tau}{2}\right) + j\phi} \frac{\sin\frac{1}{2}(\omega - \omega_c)\tau}{\omega - \omega_c} + A_0 e^{-j(\omega + \omega_c)\left(\delta + \frac{\tau}{2}\right) - j\phi} \frac{\sin\frac{1}{2}(\omega + \omega_c)\tau}{\omega + \omega_c}$$ (1.32)

Note that $S_0(\omega)$ has zero at

$$\omega - \omega_c = \frac{2n\pi}{\tau}$$ (1.33)

$$f - f = \frac{n}{\tau}, n \in I$$

1.2.2 PULSE TRAIN

Case I: Carrier phase coherent from pulse to pulse [10,22]

$$s(t) = m(t)\cos(\omega_c t + \phi)$$ (1.34)

$$m(t) = \sum_{-\infty}^{\infty} m_0(t - nT_p)$$

(1.35)

$$m_0(t) = A_0[u(t-\tau) - u(t-\tau-\tau)]$$

where T_p is the repetition period ($T_p > \tau$).

$$M(\omega) = M_0(\omega)\frac{2\pi}{T_p} \sum_{m=-\infty}^{\infty} \delta\left(\omega - \frac{2\pi m}{T_p}\right) = \frac{2\pi}{T_p} \sum_{m=-\infty}^{\infty} M_0\left(\frac{2\pi m}{T_p}\right)\delta\left(\omega - \frac{2\pi m}{T_p}\right)$$

(1.36)

Inversion of $M(\omega)$ gives

$$m(t) = \frac{1}{T_p} \sum_{m=-\infty}^{\infty} M_0\left(\frac{2m\pi}{T_p}\right)e^{j\frac{2m\pi}{T_p}t}$$

(1.37)

$$s(t) = \mathcal{R}e\left\{\frac{1}{T_p} \sum_{m=-\infty}^{\infty} M_0\left(\frac{2\pi m}{T_p}\right)\exp\left[j\left(\omega_c + \frac{2\pi m}{T_p}\right)\right]t + j\phi\right\}$$

(1.38)

Case II: Carrier phase coherent with modulating pulse [10,22]

$$s(t) = \sum_{m=-\infty}^{\infty} x_0(t - nt)$$

(1.39)

$$s(t) = \frac{1}{T_p} \sum_{m=-\infty}^{\infty} X_0\left(\frac{2\pi m}{T_p}\right)e^{j\frac{2m\pi t}{T}}$$

(1.40)

The receiver signal will be a replica of the transmitted signal, being modified by various factors, including distance, antenna, target size, change in center frequency due to target on the relative radar motion, and noise and clutter [10–15].

1.2.3 FREQUENCY-MODULATED CONTINUOUS WAVE

One fundamental issue in designing a good radar system is its capability to resolve two small targets that are located at long range with very small separation between them [1,10–15,22]. This requires a radar system to transmit a long pulse, which will have enough energy to detect a small target at long range. However, a long pulse degrades range resolution. Hence, frequency or phase modulation of the signal is used to achieve a high range resolution when a long pulse is required. The capabilities of a short-pulse and high-range-resolution radar are significant. For example, high range resolution allows resolving more than one target with good accuracy in range without using angle information. Other applications of a short-pulse and high-range-resolution radar are clutter reduction, glint reduction, multipath resolution, target classification, and Doppler tolerance. Frequency modulation (FM) is a technique where the frequency of the carrier is varied in accordance with some characteristic of the baseband modulating signal $s_b(t)$ [10]:

$$s(t) = A_0 \exp\left\{ j\left[\omega_c t + a_r \int_{-\infty}^{t} s_b(t)\,dt \right] \right\}$$ (1.41)

The instantaneous frequency of $s(t)$ can be obtained by differentiating the instantaneous phase of $s(t)$:

$$f = \frac{1}{2\pi}\frac{d}{dt}\left[\omega_c t + \beta \int_{-\infty}^{t} s_b(\tau)\,d\tau \right] = f_c + \frac{1}{2\pi}\beta s_b(t)$$ (1.42)

where f_c is the carrier frequency and β is referred to as the modulation index and is the maximum value of the phase deviation [10–12].

The linear frequency modulation (LFM) involves a transmitter frequency that is continually increasing or decreasing from a fixed reference frequency. That is, the transmitted FM signal is modified so that the frequency is modulated in a linear manner with time. That can be written as [10–12]

$$f = f_c + a_r t$$ (1.43)

$$\beta s_b(t) = 2\pi a_r t$$ (1.44)

where a_r is the frequency-changing rate or chirp rate and is mathematically defined as

$$a_r = \frac{df}{dt}$$ (1.45)

Substituting these equations into Equation 1.42, Equation 1.41 is written as

$$s(t) = A_0 \exp[j(2\pi f_c t + \pi a_r t^2)]$$ (1.46)

Figure 1.3 illustrates the basic time–frequency of an FM radar with a chirping triangle waveform, where the slope is the chirp rate. The inverse of the modulation period is the pulse repetition frequency or the modulation frequency:

$$f_m = \frac{1}{T_m} = \frac{1}{2t_0} \tag{1.47}$$

The instantaneous frequency within the modulation period is given by Equation 1.45:

$$\dot{f} = \frac{df}{dt} = \frac{\Delta f}{t_0} = 2f_m \Delta f \tag{1.48}$$

where Δf is bandwidth. The beat frequency is found by the chirp rate times the delay time:

$$f_b = \dot{f}\tau = \frac{4f_m \Delta f}{c} R \tag{1.49}$$

Note that Equation 1.49, implying the delay time from the radar to the target, $\tau = 2R/c$, is translated to the beat frequency, which is used to determine the target range. The range sensitivity is determined by the modulation bandwidth. The linear relationship between the range and the beat frequency is maintained and counted on the

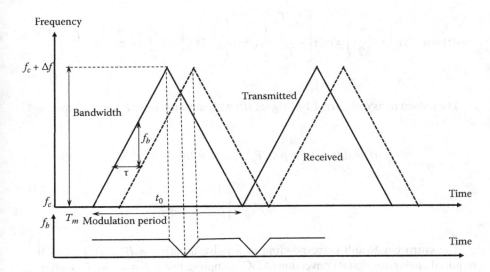

FIGURE 1.3   Triangle FMCW waveform showing generic parameters.

linearity of the chirping. The larger the bandwidth, the more difficult it is to reduce the linearity deviation during the course of chirping. Also note that resemblances of signal parameters between the pulse and frequency-modulated continuous wave (FMCW) are evident.

Two types of LFM signals are commonly used in radar applications: the pulsed LFM (also called chirp) and the continuous-wave LFM (CWLFM). In the following, the pulsed LFM signal is described.

Let $a_1(t)$ and $a_2(t)$ be defined as

$$a_1(t) = rect\left[\frac{t - \frac{T_p}{2}}{T_p}\right] \tag{1.50}$$

$$a_2(t) = A_0\,\exp[j(2\pi f_c t + \pi a_r t^2)] \tag{1.51}$$

where $rect(t)$ is a rectangular gate function, defined as

$$rect(t) = \begin{cases} 1 & \text{for } -\frac{1}{2} \le t \le \frac{1}{2} \\ 0 & \text{otherwise} \end{cases} \tag{1.52}$$

A pulsed symmetric LFM signal $p(t)$ with duration time T_p can be written as

$$s(t) = a_1(t)a_2\left(t - \frac{T_p}{2}\right) = rect\left[\frac{t - \frac{T_p}{2}}{T_p}\right] A_0 \exp\left[j\left(2\pi f_c\left(t - \frac{T_p}{2}\right) + \pi a_r\left(t - \frac{T_p}{2}\right)^2\right)\right] \tag{1.53}$$

The pulsed nonsymmetric LFM signal $s(t)$ with duration time T_p can be expressed as

$$s(t) = a_1(t)a_2(t) = p_r(t, T_p)A_0 \exp[j(2\pi f_c t + \pi a_r t^2)]$$

$$p_r(t, T_p) = \begin{cases} 1, & |t/T_p| \le 1/2 \\ 0, & \text{else} \end{cases} \tag{1.54}$$

The chirp bandwidth corresponding to a pulse duration is $B_c = |a_r| T_p$, and the required analog-to-digital conversion (ADC) sampling rate is $f_s = \alpha_{os,r} |a_r| T_p$, with $\alpha_{os,r}$ denoting an oversampling factor [3] between 1.1 and 1.4. The chirp is repeatedly transmitted with pulse repetition frequency $f_p = 1/T$; T is the period. The ratio T/T_p is called the duty cycle. For a spaceborne SAR system, the typical value is between 6 and 10, while for an airborne system it is generally chosen as 1 [3].

Figure 1.4 illustrates the envelop, phase, and spectrum of a typical transmitted signal with $T_p = 37$ μs, $a_r = 4.1 \times 10^{11}$ Hz/s, $f_s = 1.89 \times 10^7$ Hz, and $B_c = 15.17$ MHz.

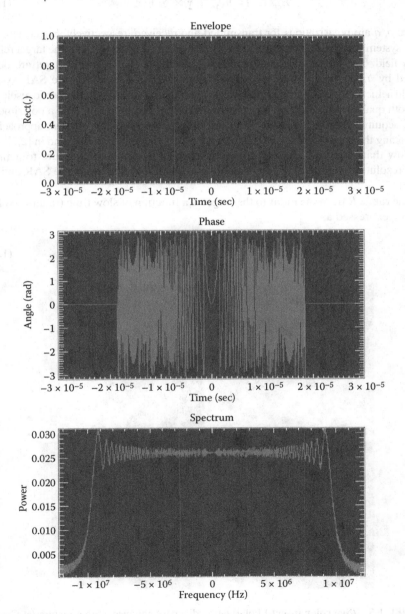

FIGURE 1.4 Typical chirp signal: envelop, phase, and spectra.

1.3 CONCEPT OF SYNTHETIC APERTURE RADAR

SAR measures the scattered fields or the scattering matrix, in both amplitude and phase, as given by [21]

$$E_{qp}(x, y) = h_{SAR}(x, y) \otimes \otimes \Gamma(x, y) \qquad (1.55)$$

where p, q are polarizations for transmitted and received, respectively, $h_{SAR}(x, y)$ is the SAR system impulse response function (see Equation 1.2), and $\Gamma(x, y)$ is the target reflectivity field. The radiometric and geometric resolutions of the sensed target are determined by $h_{SAR}(x, y)$ and the convolution operator. A good-performance SAR system should achieve high spatial resolution while maintaining sufficient radiometric resolution for both qualitative and quantitative sensing. Figure 1.5 is an example of a polarimetric SAR acquired from Radarsat-2. The polarization diversity is an efficient approach to enhancing the target features. Excellent treatment of the topic can be found in [2,7].

Now that to more concisely grasp the concept of synthetic aperture to achieve high resolution, Figure 1.6 depicts an observation relation between the SAR and the target.

The range R from the radar to the target is a function of slow time (radar traveling time) η expressed as

$$R = R(\eta) = \sqrt{R_0^2 + u^2 \eta^2} \qquad (1.56)$$

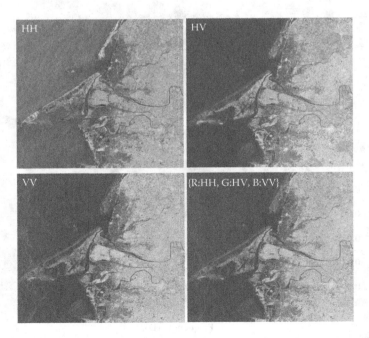

FIGURE 1.5 **(See color insert.)** Polarization diversity enhances target feature (at Canada Space Agency [CSA]).

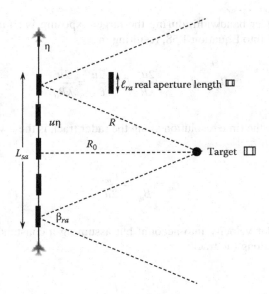

FIGURE 1.6   Concept of a synthetic aperture.

where R_0 is the closest range and u is the radar moving velocity. The phase associated with the moving radar is a function of slow time η:

$$\varphi(\eta) = -\frac{2R}{\lambda} \approx -\frac{2R_0}{\lambda} - \frac{(u\eta)^2}{\lambda R_0} \tag{1.57}$$

The induced frequency change (Doppler shift) is

$$f_d = \frac{d\varphi(\eta)}{d\eta} = -\frac{2u^2}{\lambda R_0}\eta \tag{1.58}$$

From Figure 1.5, the virtual length of the aperture subtends to the target is determined by the closest range and the real aperture beamwidth β_{ra} along the azimuth:

$$L_{sa} = \beta_{ra}R_0 = \frac{\lambda}{\ell_{ra}}R_0 \tag{1.59}$$

The time the target is exposed to the radar during the course of L_{sa}, called the target exposure time, is

$$T_a = \frac{L_{sa}}{u} = \frac{\lambda R_0}{u\ell_{ra}} \tag{1.60}$$

The total Doppler bandwidth during the target exposure is simply obtained by substituting $\eta = T_a$ into Equation 1.58, resulting in

$$B_{df} = |f_d| T_a = \frac{2u^2}{\lambda R_0} T_a = \frac{2u}{\ell_{ra}} = \frac{2u}{\lambda R_0} \tag{1.61}$$

It turns out that the time resolution along the radar track is the inverse of the total Doppler bandwidth:

$$t_{az} = \frac{1}{B_{df}} = \frac{\ell_{ra}}{2u} \tag{1.62}$$

Taking the radar velocity into account but assuming a constant, we obtain the spatial resolution along the track:

$$\delta_{az} = t_{az} u = \frac{\ell_{ra}}{2} \tag{1.63}$$

which is exactly one-half of the real aperture length; the smaller the aperture size, the finer the resolution. A great advantage of synthetic aperture is that the smaller the real aperture length, the finer is the spatial resolution along the track. In practice, the required transmitting power, imaging swath, signal-to-noise ratio, and trade-off between the imaging swath and pulse repetition frequency (PRF), among others, pose limits of the choice of the real aperture size.

The phase variations, as a function of the sensor moving velocity given in Equation 1.57, are important as far as motion compensation is concerned (see Chapter 6). If we expand the slow-time-dependent range $R(\eta)$ into Taylor series,

$$R(\eta) = R(0) + \left.\frac{\partial R}{\partial \eta}\right|_{\eta=0} \eta + \left.\frac{\partial^2 R}{\partial \eta^2}\right|_{\eta=0} \frac{\eta^2}{2} + \cdots \tag{1.64}$$

Putting the above expression into Equation 1.57, we found that

$$\varphi(\eta) = -\frac{2R}{\lambda} = -\frac{2R_0}{\lambda} - \frac{2}{\lambda} u\eta \cos\alpha + \frac{(u\eta)^2}{2\lambda R_0} \sin^2\alpha + \cdots \tag{1.65}$$

where α is the angle between the velocity vector and slant range. $u\cos\alpha$ is commonly termed the radial velocity. The first term on the right-hand side of Equation 1.65 is a constant phase, while the second term is a linear phase. The third term is a quadratic phase varying as a squared velocity. It is this phase term for which the motion compensation is a major concern. Referring to Figure 1.6, the maximum quadratic phase occurs at

$$\varphi_q\left(\frac{T_a}{2}\right) = \frac{(u\eta)^2}{2\lambda R_0}\sin^2\alpha\Big|_{\eta=T_a/2} = \frac{u^2}{8\lambda R_0}T_a^2 \tag{1.66}$$

From Equations 1.60 and 1.63, we can express Equation 1.66 as a function of spatial resolution:

$$\varphi_{q,\max} = \frac{u^2}{8\lambda R_0}T_a^2 = \frac{\lambda}{16}\frac{R_0}{\delta_{az}^2} \tag{1.67}$$

We see from the above relation that the higher the resolution, the larger the quadratic phase that is induced—posing a challenge work for focusing well.

APPENDIX: USEFUL FOURIER TRANSFORM PAIRS AND PROPERTIES

This appendix gives some basic functions and their Fourier transforms [19–21]. These functions are fundamental in designing the radar signal. Because SAR is dealing with signals in the time–frequency domain, the transform pair should find itself useful throughout this book. Also included are basic signal operations, such as correlation, convolution, modulation–demodulation, and the Hilbert transform.

1. *Pulse (rectangular) function*

$$p(t) = \begin{cases} 1, & |t| \le \tau \\ 0, & |t| > \tau \end{cases} \tag{A.1}$$

$$p(t) \Leftrightarrow 2\tau\,\text{sinc}(\omega t) \tag{A.2}$$

2. *Cosine function*

$$\cos(\omega_0 t) = \frac{e^{j\omega_0 t} + e^{-j\omega_0 t}}{2} \tag{A.3}$$

$$\cos(\omega_0 t) \Leftrightarrow \pi[\delta(\omega - \omega_0) + \delta(\omega + \omega_0)] \tag{A.4}$$

3. *Signum function*

$$\text{sgn}(t) = \begin{cases} 1, & t > 0 \\ -1, & t < 0 \end{cases} \tag{A.5}$$

$$\text{sgn}(t) \Leftrightarrow \frac{2}{j\omega} \tag{A.6}$$

From the duality, it is noted that

$$\frac{1}{t} \Leftrightarrow -j\pi\,\text{sgn}(\omega) \tag{A.7}$$

4. *Heaviside step function*

$$u(t) = \begin{cases} 1, & t > 0 \\ 0, & t < 0 \end{cases} \tag{A.8}$$

$$u(t) \Leftrightarrow \delta(\omega) + \frac{1}{j\omega} \tag{A.9}$$

Note that the pulse may be represented by the step function as $p(t, \tau) = u(t) - u(t - \tau)$.

5. *Convolution*

$$f \otimes g = \int_{-\infty}^{\infty} f(t)g(\tau - t)\,dt \tag{A.10}$$

$$f(t) \otimes g(t) \Leftrightarrow F(\omega)G(\omega) \tag{A.11}$$

6. *Correlation*

$$f \cdot g = \int_{-a}^{\infty} f(t)g(t - \tau)\,dt, \quad t \in \mathbb{R} \tag{A.12}$$

$$f(t) \cdot g(t) \Leftrightarrow F(-\omega)G(\omega), \quad t \in \mathbb{R} \tag{A.13}$$

7. *Modulation and demodulation*

$$f(t)\cos(\omega_c t) \Leftrightarrow F(\omega) \otimes \pi[\delta(\omega - \omega_c) + \delta(\omega + \omega_c)] = \pi[F(\omega - \omega_c) + F(\omega + \omega_c)] \tag{A.14}$$

$$f(t)\cos(\omega_c t)\cos(\omega_c t) \Leftrightarrow \pi^2[2F(\omega) + F(\omega - 2\omega_c) + F(\omega + 2\omega_c)] \tag{A.15}$$

8. *Hilbert transform and quadrature detection*

$$q(t) = \hat{H}[f(t)] \triangleq \frac{1}{\pi} \int_{-\infty}^{\infty} \frac{f(\tau)}{t - \tau} d\tau = \frac{1}{\pi t} \otimes f(t) \tag{A.16}$$

$$Q(\omega) = \mathcal{F}[q(t)] = [-j\,\text{sgn}(\omega)][F(\omega)] \tag{A.17}$$

$$q(t) = \mathcal{F}^{-1}[-j\,\text{sgn}(\omega)F(\omega)] \tag{A.18}$$

It is easily seen that $q(t)$ and $f(t)$ have a phase difference of 90°. As an example, if $f(t) = y(t)\cos(\omega_c t)$, $\omega_c > B$ and $y(t)$ has spectrum $Y(\omega)$, $|\omega| \le B$, then $q(t) = \hat{H}[f(t)] = y(t)\sin(\omega_c t)$; namely, $q(t)$ and $f(t)$ form an orthogonal in-phase and quadrature-phase (I-Q) pair.

REFERENCES

1. Barton, D. K., *Modern Radar System Analysis*, Artech House, Norwood, MA, 1988.
2. Cloude, S. R., *Polarisation: Applications in Remote Sensing*, Oxford University Press, Oxford, 2009.
3. Cumming, I. G., and Wong, F. H., *Digital Processing of Synthetic Aperture Radar Data: Algorithms and Implementation*, Artech House, Norwood, MA, 2005.
4. Curlander, J. C., and McDonough, R. N., *Synthetic Aperture Radar: Systems and Signal Processing*, Wiley-Interscience, New York, 1991.
5. Elachi, C., *Space-Borne Radar Remote Sensing: Applications and Techniques*, IEEE Press, New York, 1988.
6. Franceschitti, G., and Lanari, R., *Synthetic Aperture Radar Processing*, CRC Press, New York, 1999.
7. Lee, J. S., and Pottier, E., *Polarimetric Radar Imaging: From Basics to Applications*, CRC Press, New York, 2009.
8. Maitre, H., *Processing of Synthetic Aperture Radar (SAR) Images*, Wiley-ISTE, New York, 2008.
9. Oliver, C., and Quegan, S., *Understanding Synthetic Aperture Radar Images*, SciTech Publishing, Raleigh, NC, 2004.
10. Richards, M. A., *Fundamentals of Radar Signal Processing*, 2nd ed., McGraw-Hill, New York, 2014.
11. Skolnik, M., *Introduction to Radar Systems*, McGraw-Hill, New York, 2002.
12. Skolinik, M. I., ed., *Radar Handbook*, 3rd ed., McGraw-Hill, New York, 2008.
13. Soumekh, M., *Synthetic Aperture Radar Processing*, John Wiley & Sons, New York, 1999.
14. Sullivan, R. J., *Microwave Radar: Imaging and Advanced Concepts*, Artech House, Norwood, MA, 2000.
15. Ulaby, F. T., Moore, R. K., and Fung, A. K., *Microwave Remote Sensing: Active and Passive*, vol. 2, *Radar Remote Sensing and Surface Scattering and Emission Theory*, Artech House, Norwood, MA, 1982.
16. Ulaby, F. T., and Elachi, C., eds., *Radar Polarimetry for Geoscience Applications*, Artech House, Norwood, MA, 1990.
17. Lathi, B. P., *Signal Processing and Linear Systems*, Oxford University Press, New York, 2000.

18. Etten, W. van, *Introduction to Random Signals and Noise*, John Wiley & Sons, New York, 2005.
19. Bracewell, R. M., *The Fourier Transform and Its Applications*, McGraw-Hill, New York, 1999.
20. Papoulis, A., *Fourier Integral and Its Applications*, McGraw-Hill, New York, 1962.
21. Blackledge, J. M., *Quantitative Coherent Imaging: Theory, Methods and Some Applications*, Elsevier, Netherlands, 2012.
22. Berkowitz, R. S., ed., *Modern Radar: Analysis, Evaluation, and System Design*, John Wiley & Sons, New York, 1965.
23. Davenport, W. B., Jr., and Root, W. L., *An Introduction to the Theory of Random Signals and Noise*, Wiley-IEEE Press, New York, 1987.

2 SAR Models

2.1 INTRODUCTION

Synthetic aperture radar (SAR) is a complex system that integrates two major parts: data collector and image formatter. In the stage of data collection, a radar transmits electromagnetic waves toward the target and receives the scattered waves [1–21]. The transmitted signal can be modulated into certain types, commonly linearly frequency modulated with pulse or continuous waveform, with or without coding [22]. The process involves signal transmission from the generator, through various types of guided devices, to the antenna, by which the signal is radiated into free space and then undergoes propagation. During the course of propagation, the signal may be attenuated before and after impinging upon the targets, experiencing certain degrees of absorption and scattering. The measured scattered signal made in bistatic or monostatic configurations is essentially in the time–frequency (delay time–Doppler frequency) domain. The role of image formatter is then to map the time–frequency data into the spatial domain in which the targets are located [5,15,16,23–29]. The mapping from the data domain to the image domain, and eventually into the target or object domain, must minimize both geometric and radiometric distortions. This chapter provides an overview of two models that define the SAR operational process–physical model and system model–to facilitate our discussions in Chapters 5, 7, and 8.

2.2 PHYSICAL MODEL

Understanding the electromagnetic wave interactions with the target of interest is essential to derive the target information from the measured electric field [6,8,17–21,30–32]. In this section, we present the radar speckle properties embedded in the scattering process. Unlike common practice, speckle statistics are derived from radar signal aspects. Here we show them from the wave scattering and propagation points of view. For purposes of demonstration, only rough surface scattering is shown. Other scattering mechanisms outlined in Section 2.3.2 can be done equally well. As mentioned earlier, the scattering matrix contains more complete information of the target's geophysical properties that relate to the permittivity and conductivity. Physically, SAR measures the vector electric field **E**, governed by the following wave equation [3,32]:

$$\nabla \times \nabla \times \mathbf{E}(\vec{r}) + j\omega\mu(\sigma + j\omega\varepsilon)\mathbf{E}(\vec{r}) = 0 \tag{2.1}$$

or equivalently

$$(\nabla^2 + k^2)\mathbf{E} = k^2(1 - \varepsilon_r)\mathbf{E} + jk\eta_0\sigma\mathbf{E} - \nabla(\mathbf{E} \cdot \nabla \ln \varepsilon) \tag{2.2}$$

19

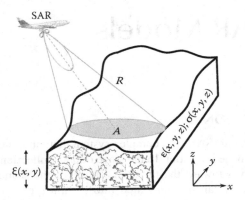

FIGURE 2.1 SAR observation of a physical target.

where the target is characterized by permittivity $\varepsilon(\vec{r})$ and conductivity $\sigma(\vec{r})$ with random height $z = \xi(x, y)$], as shown in Figure 2.1 (refer to [3]). Note that the random height, such as a rough surface, top of a vegetation canopy, and snow layers, to name a few, should be electromagnetically interpreted in terms of radar wavelength.

Keep in mind that the dielectric spectra of ε, σ are related to the geophysical parameters of interest, achieved by means of radar or radiometric remote sensing. It is possible to infer the unknown geophysical parameters by effectively collecting the angular, frequency, and polarization responses and their correlations, such as configurations of interferometry and tomography. Depending on the exploring wavelength, SAR obtains the ε, σ map at a certain penetration height $\xi(x, y) =$ constant. That is, only a two-dimensional (2D) reflectivity map is available if only a narrow bandwidth is utilized. The compression of the three-dimensional (3D) $\varepsilon(x, y, z)$, $\sigma(x, y, z)$ to 2D $\varepsilon(x, y)$, $\sigma(x, y)_{z=\text{constant}}$ results in the loss of height information and is mostly nonreversible unless others means are devised. SAR tomography offers potential reconstruction of the height profile, but that is beyond the scope of this book.

2.2.1 RADAR SCATTERING POWER AND SCATTERING COEFFICIENT

The total electric field **E** in Equation 2.1 is the sum of the incident field $\mathbf{E}^i$, which is known, and the scattered field $\mathbf{E}^s$, which to a certain extent carries the target information and is to be measured. The scattered field is random in nature in the presence of random media. In what follows, we use rough surface scattering to illustrate the computation of scattering power, scattering coefficient, and later the radar speckle from electromagnetic fields of point of view. Figure 2.2 depicts the geometry of radar scattering in the spherical coordinate system, where $\hat{k}_i$, $\hat{k}_s$ are incident and scattering direction vectors, respectively.

$$\hat{k}_i(\theta_i, \phi_i) = \hat{x}\sin\theta_i\cos\phi_i + \hat{y}\sin\theta_i\sin\phi_i - \hat{z}\cos\theta_i \tag{2.3}$$

$$\hat{k}_s(\theta_s, \phi_s) = \hat{x}\sin\theta_s\cos\phi_s + \hat{y}\sin\theta_s\sin\phi_s + \hat{z}\cos\theta_s \tag{2.4}$$

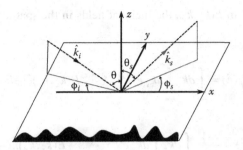

FIGURE 2.2 Geometry of radar scattering.

Note that the scattering geometry can be described in a wave system, the forward scattering alignment (FSA), or in an antenna system, the backward scattering alignment (BSA). Care must be taken when specifying the scattering geometry. The conversion between FSA and BSA is straightforward [6,9,20,21].

The scattered field can be related to the incident field through the scattering matrix **S**, which describes the target characteristics:

$$\mathbf{E}^s = \frac{e^{jkR}}{R}\mathbf{S}\left(\hat{k}_i,\hat{k}_s\right)\mathbf{E}^i \tag{2.5}$$

where the incident and scattered wave vectors, $\hat{k}_i$, $\hat{k}_s$, are defined in the spherical coordinate, k is the wavenumber, and R is the range from the radar to the target:

$$\mathbf{S} = \begin{bmatrix} S_{hh} & S_{hv} \\ S_{vh} & S_{vv} \end{bmatrix} \tag{2.6}$$

As an example, let's consider the radar scattering from a homogeneous surface, as shown in Figure 2.2. The spectrum of the incident wave $E(k_x, k_y)$ is [32–37]

$$E(k_x,k_y) = \frac{1}{4\pi^2}\int\limits_{-\infty}^{\infty} dx \int\limits_{-\infty}^{\infty} dy\, e^{-j(k_x x + k_y y)} e^{j(k_{ix}x + k_{iy}y)} g_t(\theta_i,\phi_i) \tag{2.7}$$

where k is the wavenumber of free space, $k_{ix} = k \sin\theta_i \cos\phi_i$; $k_{iy} = k \sin\theta_i \sin\phi_i$. The inclusion of the transmitting antenna gain pattern g_t in the Equation 2.7 ensures that the incident wave is zero at the edge of the surface being illuminated, and thus no diffraction scattering occurs. In the spectral domain (k_x, k_y), the incident wave is centered at the incidence angle and decreases quickly away from the incidence angle.

Given the spectrum $E(k_x, k_y)$, the incident fields in the spatial domain are calculated by

$$\mathbf{E}_p^i(\vec{r}) = \int\limits_{-\infty}^{\infty} dk_x \int\limits_{-\infty}^{\infty} dk_y \, e^{j(k_x x + k_y y - k_z z)} E(k_x, k_y) \hat{e}(-k_z) \tag{2.8}$$

$$\mathbf{H}_q^i(\vec{r}) = -\frac{1}{\eta_0} \int\limits_{-\infty}^{\infty} dk_x \int\limits_{-\infty}^{\infty} dk_y \, e^{j(k_x x + k_y y - k_z z)} E(k_x, k_y) \hat{h}(-k_z) \tag{2.9}$$

where η_0 is the wave impedance of free space. The unit vectors $\hat{e}(-k_z)$ and $\hat{h}(-k_z)$ are

$$\hat{e}(-k_z) = \frac{1}{k_\rho} (\hat{x} k_y - \hat{y} k_x) \tag{2.10}$$

$$\hat{h}(-k_z) = \frac{k_z}{k k_\rho} (\hat{x} k_x + \hat{y} k_y) + \hat{z} \frac{k_\rho}{k} \tag{2.11}$$

for the horizontal polarization wave incidence and

$$\hat{e}(-k_z) = \frac{k_z}{k k_\rho} (\hat{x} k_x + \hat{y} k_y) + \hat{z} \frac{k_\rho}{k} \tag{2.12}$$

$$\hat{h}(-k_z) = -\frac{1}{k_\rho} (\hat{x} k_y - \hat{y} k_x) \tag{2.13}$$

for the vertical polarization wave incidence. In the Equation 2.13

$$k_\rho = \sqrt{k_x^2 + k_y^2}$$

and

$$k_z = \begin{cases} \sqrt{k^2 - k_\rho^2}, & k_\rho < k \\ j\sqrt{k_\rho^2 - k^2}, & k_\rho > k \end{cases} \tag{2.14}$$

The incident power of p polarization is P_p^i

$$P_p^i = \frac{1}{2} \int \mathcal{R}_e \left[\mathbf{E}_p^i(\vec{r}) \times \mathbf{H}_p^{i*}(\vec{r}) \right] \cdot \hat{n} \, dA \qquad (2.15)$$

where $\hat{n}$ is the normal vector of the rough surface (see Figures 2.1 and 2.2).

Consider vertical polarized incidence at incident angle (θ_i, ϕ_i); the method of moment (MoM) solution of equations gives the tangential surface fields $\mathbf{J}_v(\theta_i, \phi_i)$ and $\mathbf{M}_v(\theta_i, \phi_i)$ for incident vertical polarization. Then the vertical polarized and horizontal polarized bistatic fields are calculated by integration of the tangential surface fields over the area $L_s \times L_s$:

$$E_{vv}^s(\theta_s, \phi_s; \theta_i, \phi_i) = \frac{jk}{4\pi} \int_A dA \left[\hat{v}_s \cdot \eta_0 \mathbf{J}_v(\theta_i, \phi_i) + \hat{h}_s \cdot \mathbf{M}_v(\theta_i, \phi_i) \right] e^{-jk\hat{k}_s \cdot \vec{r}} \qquad (2.16)$$

$$E_{hv}^s(\theta_s, \phi_s; \theta_i, \phi_i) = \frac{jk}{4\pi} \int_A dA \left[\hat{h}_s \cdot \eta_0 \mathbf{J}_v(\theta_i, \phi_i) - \hat{v}_s \cdot \mathbf{M}_v(\theta_i, \phi_i) \right] e^{-jk\hat{k}_s \cdot \vec{r}} \qquad (2.17)$$

where $\hat{v}_s = \hat{x} \cos\theta_s \cos\phi_s + y \cos\theta_s \sin\phi_s - \hat{z} \sin\theta_s$ and $\hat{h}_s = -\hat{x} \sin\phi_s + y \cos\phi_s$ denote vertical and horizontal polarizations. Consider horizontal polarized incidence; solving Maxwell equations gives tangential fields $\mathbf{J}_h(\theta_i, \phi_i)$ and $\mathbf{M}_h(\theta_i, \phi_i)$. Then the vertical polarized and horizontal polarized bistatic fields are given respectively by [32,34,35]

$$E_{vh}^s(\theta_s, \phi_s; \theta_i, \phi_i) = \frac{jk}{4\pi} \int_A dA \left[\hat{v}_s \cdot \eta_0 \mathbf{J}_h(\theta_i, \phi_i) + \hat{h}_s \cdot \mathbf{M}_h(\theta_i, \phi_i) \right] e^{-jk\hat{k}_s \cdot \vec{r}} \qquad (2.18)$$

$$E_{hh}^s(\theta_s, \phi_s; \theta_i, \phi_i) = \frac{jk}{4\pi} \int_A dA \left[\hat{h}_s \cdot \eta_0 \mathbf{J}_h(\theta_i, \phi_i) + \hat{h}_s \cdot \mathbf{M}_h(\theta_i, \phi_i) \right] e^{-jk\hat{k}_s \cdot \vec{r}} \qquad (2.19)$$

The bistatic coefficients and backscattering coefficients are normalized by the incident power.

$$\gamma_{qp}(\theta_s, \phi_s; \theta_i, \phi_i) = \frac{1}{N} \frac{1}{2\eta_0 P_p^i} \sum_{n=1}^{N} \left| E_{qp,n} \theta_s, \phi_s; \theta_i, \phi_i) \right|^2 \qquad (2.20)$$

The scattering coefficients are

$$\sigma_{qp}^o(\theta_s, \phi_s; \theta_i, \phi_i) = \cos\theta_i \gamma_{qp}(\theta_s, \phi_s; \theta_i, \phi_i) \qquad (2.21)$$

For polarimetric studies, we calculate the normalized scattering matrix for each nth realization by

$$S_{qp} = E_{qp}\sqrt{\frac{\cos\theta_i}{2\eta_0 P_p^i}} \tag{2.22}$$

The radar cross section can be obtained in terms of the scattering matrix

$$\sigma = 4\pi\left|\mathbf{S}\left(\hat{k}_i,\hat{k}_s\right)\right|^2 \tag{2.23}$$

From the radar equation, the radar-received power from such a target is

$$P_r = \frac{P_t g_t(\theta_i,\phi_i)g_r(\theta_s,\phi_s)\lambda^2\sigma(\theta_i,\phi_i;\theta_s,\phi_s)}{(4\pi)^3 R_r^2 R_t^2} \tag{2.24}$$

where g_t, g_r denote the transmitting and receiving antenna patterns, respectively. For a distributed target, the scattering coefficient is of interest, which is the radar cross section normalized by the antenna illuminating area A (Figure 2.1).

$$\sigma^o(\theta_s,\phi_s;\theta_i,\phi_i) = \frac{\langle\sigma\rangle}{A} = \frac{4\pi}{A}\left|\mathbf{S}(\hat{k}_i,\hat{k}_s)\right|^2 \tag{2.25}$$

The received power from a distributed or extended target is

$$P_r = \iint\frac{P_t g_t(\theta_i,\phi_i)g_r(\theta_s,\phi_s)\lambda^2\sigma^o(\theta_i,\phi_i;\theta_s,\phi_s)}{(4\pi)^3 R_r^2 R_t^2}\,dA \tag{2.26}$$

2.2.2 Antenna Illuminated Area

In estimating the radar cross and scattering coefficients, as given in Equations 2.25 and 2.26, it is necessary to know the antenna footprint or the illuminated area A. At backscattering, the transmitting and receiving antenna patterns projected onto the target that they extended is assumed to be completely overlaid. This is true when one antenna is used for both transmitting and receiving. For a bistatic or multiple-input multiple-output (MIMO) SAR system, the common illuminated area confined by the transmitting and receiving antenna should be considered for more precise estimation of the scattering coefficient.

2.2.3 Scattering Effects

Wave scattering from the SAR target, be it point or distributed, is a complex process. Generally, the scattering can be volume scattering, surface scattering, dihedral scattering, or a combination of these. Because of the coherent process, inherently, separation of one scattering mechanism from another is difficult. Nevertheless, efforts have been devoted to interpreting these scattering effects in order to infer a target's geometric and radiometric information. Target information retrieved from polarimetric synthetic aperture radar (PolSAR) images has been studied extensively since the 1980s [6,8,9,21]. The developed target decomposition theorems [6] opened a new era for radar polarimetry and have made PolSAR more powerful than ever before, and thus of practical application for the remote sensing of terrains. Excellent treatment of these subjects from fundamentals to applications may be found in [6,9] and the relevant references cited therein. In the development of various types of decomposition, it has been realized that five physical mechanisms are involved in radar scattering from targets: rough surface scattering, dihedral reflection or low-order multiple scattering, volume scattering, surface scattering attenuated by the presence of volume targets, and scattering from anisotropic targets such as tree trunks or large branches. It is apparent that surface scattering from the ground underneath the volume targets is one of the most important scattering mechanisms. A thorough review and unification of target decomposition is provided by Cloude [6], where simplified models are used to characterize these five scattering mechanisms:

1. Backscattering from a rough surface
2. Low-order multiple scattering, as occurs from dihedral effects in forest and urban areas
3. Random volume backscatter from a nonpenetrable layer of discrete scatterers
4. Surface scattering after propagation through a random medium
5. Single scattering from anisotropic structures such as tree trunks

Figure 2.3 is an example to illustrate the scattering effects by rough classification. The SAR image was acquired by the NASA-JPL AIRSAR system at the fully

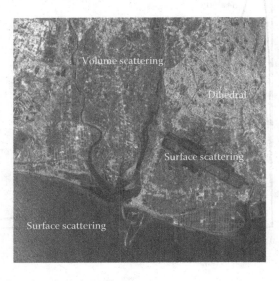

FIGURE 2.3 Basic radar scattering effects.

polarimetric L-band. City blocks have the dihedral corner reflector (double-bounce) type of scattering. Surface scattering dominates on ocean, desert, and bare surfaces (e.g., runways). Forested and vegetated areas present volume scattering.

2.2.4 RADAR SPECKLES

For fully developed speckles, the amplitude of the returns A follows the Rayleigh distribution [9,10,27]:

$$p_A(A \mid \sigma^2) = \frac{2A}{\sigma^2} \exp\left(-\frac{A^2}{\sigma^2}\right), \quad A \geq 0 \qquad (2.27)$$

where σ^2 is variance and can be seen as the homogeneous target reflectivity within a resolution cell or pixel. Figure 2.4 displays the Rayleigh distribution for various values of σ^2.

A measure of speckle strength is the coefficient of variation or contrast, defined as

$$\gamma_a = \frac{\langle A^2 \rangle - \langle A \rangle^2}{\langle A \rangle^2} = \sqrt{\frac{4}{\pi}} - 1 \approx 0.5227 \qquad (2.28)$$

It is a constant for a Rayleigh distribution, which is a single-parameter model.

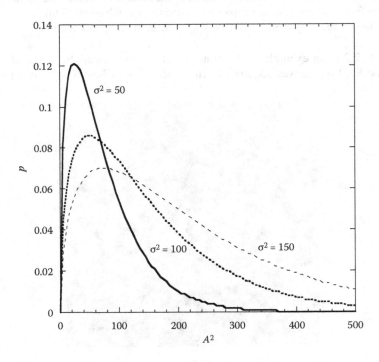

FIGURE 2.4 Rayleigh distribution for different values of $\sigma^2 = 50, 100, 150$.

The intensity $I = A^2$ has exponential or Laplace distribution:

$$p_I(I \mid \sigma^2) = \frac{1}{\sigma^2} \exp\left(-\frac{I}{\sigma^2}\right), \quad I \geq 0 \tag{2.29}$$

Figure 2.5 shows the intensity distribution for $\sigma^2 = 50, 100, 150$ (see [27]). The coefficient of variation for the intensity is defined as

$$\gamma_I = \frac{\langle I^2 \rangle - \langle I \rangle^2}{\langle I \rangle^2} = 1 \tag{2.30}$$

The inherent contrast is a constant of 1. For a fully developed speckle, the contrast can be regarded as the signal-to-noise ratio and constant. For this reason, the speckle is multiplicative. That is, the only way to increase the contrast is to reduce the variance, normally by means of multilooking. Before proceeding, we display the amplitude, intensity image, and a coefficient of variation, calculated with a moving window of size 7×7 pixels, as shown in Figure 2.6.

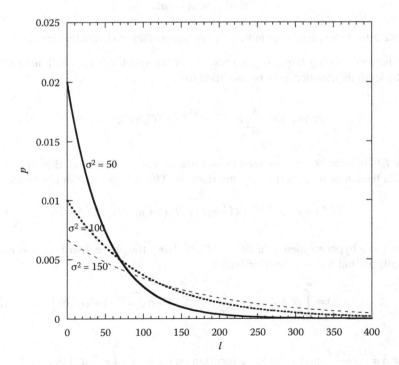

FIGURE 2.5   Exponential distribution for different values of $\sigma^2 = 50, 100, 150$.

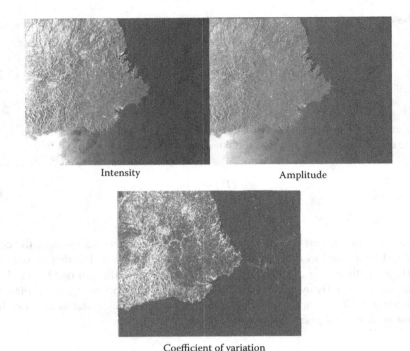

Intensity   Amplitude

Coefficient of variation

FIGURE 2.6 Example of amplitude, intensity, and coefficient of variation images.

If there is a strong target in the presence of the speckle field with amplitude C_0, the Rayleigh distribution may be modified to

$$p(A/\sigma^2) = \frac{A}{\pi\sigma^2} e^{-\left(A^2+C_0^2\right)/\sigma^2} I_0(2AC_0/\sigma^2), \quad A \geq 0 \tag{2.31}$$

where $I_0(\cdot)$ is a zero-order modified Bessel function of the first kind. Equation 2.31 is called a Rician or Rice–Nakagami distribution. The mth moment of amplitude A is

$$\langle A^m \rangle = \sigma^m e^{-2C_0^2/\sigma^2} \Gamma(1+m/2) \, _1F_1\left(1+m/2, 1, 2C_0^2/\sigma^2\right) \tag{2.32}$$

where F is a hypergeometric function [38,39]. Now, the phase distribution is no longer uniform, but follows the distribution

$$p(\phi) = \int_0^\infty p(A,\phi)\, dA = \frac{1}{2\pi} e^{-C_0^2/\sigma^2} \left[1 + \delta\sqrt{\pi} e^{G^2}\left(1 + erf(\delta)\right)\right] \tag{2.33}$$

where $\delta = \dfrac{C_0 \cos\phi}{\sigma}$ and erf is error function $erf(z) = \dfrac{2}{\sqrt{\pi}} \displaystyle\int_0^z e^{-t^2}\, dt$ [38].

To reduce the variance, the multilooking process is applied:

$$I = \frac{1}{L} \sum_{k=1}^{L} I_k \tag{2.34}$$

where I_k is an individual single look intensity image and L is the number of looks. If L-looks are independent and have equal power, then [10,27]

$$p_I(I \mid \sigma^2) = \left(\frac{L}{\sigma^2} \right)^L \frac{1}{\Gamma(L)} \exp\left(-\frac{LI}{\sigma^2} \right) I^{L-1}, \quad I \ge 0 \tag{2.35}$$

which has a gamma distribution with a degree of freedom $2L$. It is straightforward to verify that the speckle strength

$$\gamma_s = \frac{\langle I^2 \rangle - \langle I \rangle^2}{\langle I \rangle^2} = \frac{1}{L} \tag{2.36}$$

Compared to Equation 2.30, it is reduced by a factor of L at the cost of coarse spatial resolution. The L-look amplitude has distribution

$$p_A(A \mid \sigma^2) = 2 \left(\frac{L}{\sigma^2} \right)^L \frac{A^{(2L-1)}}{\Gamma(L)} \exp\left(-\frac{LA^2}{\sigma^2} \right), \quad A \ge 0 \tag{2.37}$$

The appendix gives the proof of Equations 2.35 and 2.37. Figure 2.7 displays the L-look amplitude distribution for $L = 1, 4, 8, 32, 128$. The effect of L-looking to reduce the speckle variance and increase the accuracy of the radar cross section (RCS) estimate is clear.

In practice, these L-look images are correlated within the total system bandwidth because subbands are practically correlated with each other. The correlation matrix between L-looks is expressed as [27]

$$\mathbf{C}_v = \sigma^2 \begin{bmatrix} 1 & \rho_{12} & \cdots & \rho_{1L} \\ \rho_{21}^* & 1 & & \\ \cdot & & \cdot & \\ \cdot & & \cdot & \cdot \\ \rho_{L1}^* & \cdot & \cdot & 1 \end{bmatrix} \tag{2.38}$$

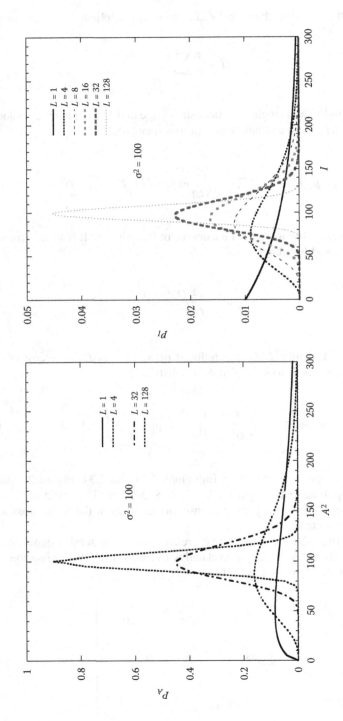

FIGURE 2.7 *L*-look amplitude and intensity distributions for *L* = 1, 4, 8, 16, 32, 128 with σ² = 100.

where * is complex conjugate. If $\lambda'_k = \lambda_k/\sigma^2$, $k = 1, 2, ..., L$, are real eigenvalues of the correlation matrix, then the L-look correlated intensity image has distribution

$$p_I(I \mid \sigma^2) = \sum_{k=1}^{L} \frac{L}{\sigma^2 \lambda'_k \prod_{j \neq k}^{L} (1 - \lambda'_j/\lambda'_k)} e^{-\frac{LI}{\sigma^2 \lambda'_k}}, \quad I \geq 0; k = 1, 2, ..., L, \quad (2.39)$$

Taking the correlation between looks, the effective number of looks is

$$L' = \frac{L}{1 + \frac{2}{L} \sum_{m=1}^{L-1} \sum_{n=m+1}^{L} \rho_{mn}}, \quad L' \leq L \quad (2.40)$$

The correlation coefficient ρ is usually unknown and may be estimated from the distribution or from real data over a uniform, homogeneous area.

The speckle model presented so far is all about being featureless, that is, the homogeneous illuminated area. For heterogeneous areas, as in many practical situations, extensive studies [27,33,40,41] suggest that K-distribution is the most reasonable clutter model. The model is proposed based on the wave scattering process. Other models are also proposed for specific types of target. For example, for surface-like clutter such as vegetation canopy and airport runways, the radar signal statistics follow a fairly well Weibull distribution or gamma distribution [13,14]. Recall that the speckle or clutter distribution is dependent to a certain pixel size.

For a gamma-distributed RCS, the L-look amplitude distribution is [9,10]

$$p_A(A) = \int p_A(A \mid \sigma^2) p(\sigma^2) d\sigma^2 = \frac{4(L\alpha)^{(\alpha+L)/2}}{\Gamma(L)\Gamma(\alpha)} A^{(\alpha+L)-1} K_{\alpha-L}\left(2A\sqrt{\alpha L}\right), \quad A \geq 0$$

$$(2.41)$$

where $K_n(\cdot)$ is the modified Bessel function of the second kind, ordered n, and v is an order parameter. When $\alpha \to \infty$, we approach the Rayleigh case; that is, it tends to be more homogeneous and contains pure speckle (Figure 2.8). Notice that the mth moment of the amplitude is

$$\langle A^m \rangle = \int A^m p_A(A) dA = \frac{\Gamma(L + m/2)\Gamma(\alpha + m)}{L^{m/2}\Gamma(L)\alpha^{m/2}\Gamma(\alpha)} \quad (2.42)$$

Using the above formula, we can estimate the order parameter from the observed data and test the goodness of fit.

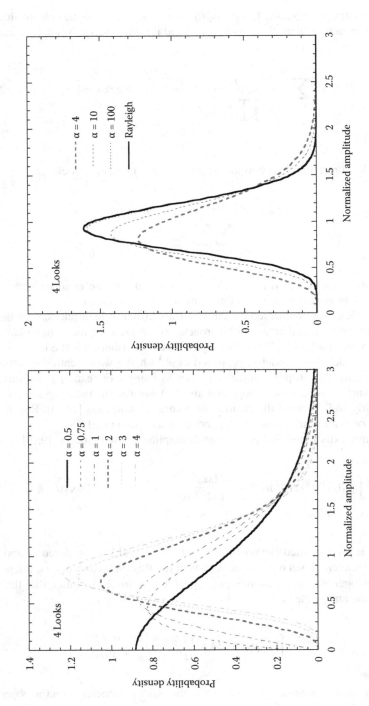

FIGURE 2.8 *K*-Distribution of *L*-look amplitude at different order parameters. When α → ∞, it approaches the Rayleigh model, as seen from the plot.

Based on circular Gaussian statistics, Lee and coworkers [9,42] derived the following n-look distributions of phase difference ψ between xx and yy polarizations:

$$p(\psi) = \frac{\Gamma(L+\frac{1}{2})\left(1-|\rho|^2\right)^L \beta}{2\sqrt{\pi}\Gamma(L)(1-\beta^2)^{L+(1/2)}} + \frac{\left(1-|\rho|^2\right)^L}{2\pi} F(L,1;\tfrac{1}{2};\beta^2), \quad -\pi < (\psi - \vartheta) \le \pi$$

(2.43)

where $\beta = |\rho|\cos(\psi - \vartheta)$, $\rho = \dfrac{\langle S_{xx}S_{yy}^*\rangle}{\sqrt{\langle|S_{xx}S_{xx}^*|\rangle\langle|S_{yy}S_{yy}^*|\rangle}} = |\rho|e^{j\vartheta}$, $F(\cdot)$ is a Gaussian hyper-

geometric function, and L is the number of looks. The distribution of the phase ψ given at Equation 2.43 is symmetrical about ϑ, ϑ is the mean with modulus 2π, and the standard deviation of ψ is independent of ϑ. Following [42], the statistical distribution of the amplitude ratio can be written as

$$p^{(L)}(v) = \frac{2\varsigma^L\Gamma(2L)\left(1-|\rho_c|^2\right)^L(\varsigma+z^2)v^{2L-1}}{\Gamma(L)\Gamma(L)\left[(\varsigma+v^2)^2 - 4\varsigma|\rho_c|^2 v^2\right]^{(2L+1)/2}}$$

(2.44)

where v is the amplitude ratio between two polarized channels xx and yy, and $\varsigma = \dfrac{C_{xx}}{C_{yy}}$ [9,42].

To illustrate the validity of the Rayleigh distribution for a homogeneous target, we generate a realization of random rough surfaces $\xi(x, y)$ with the heights varying in both horizontal directions. For each nth realization, we solve the Maxwell equation to calculate the bistatic field $\mathbf{E}_n^s$ according to Equations 2.16 through 2.19. For incident polarization p and scattered polarization q, the scattered field is denoted as $\mathbf{E}_{qp,n}(\theta_s, \phi_s)$. These are calculated for a total of N realizations. The scattered field can be decomposed into coherent and incoherent fields. The coherent scattered field $\langle \mathbf{E}^s(\theta_s, \phi_s)\rangle$ is calculated by averaging over realizations:

$$\langle \mathbf{E}^s(\theta_s, \phi_s)\rangle = \frac{1}{N}\sum_{n=1}^{N}\mathbf{E}_n^s(\theta_s, \phi_s)$$

(2.45)

where $\mathbf{E}_n^s(\theta_s, \phi_s)$ is the scattered field for realization n, and angular brackets represent averaging over realizations. The incoherent scattered field for the realization n is expressed as

$$\mathbf{E}_n^{s,incoh}(\theta_s, \phi_s) = \mathbf{E}_n^s(\theta_s, \phi_s - \langle \mathbf{E}_n^s(\theta_s, \phi_s\rangle$$

(2.46)

For a surface of infinite extent, the coherent field exists only in the specular direction and will not contribute to radar backscattering. Thus, for a very large surface, only

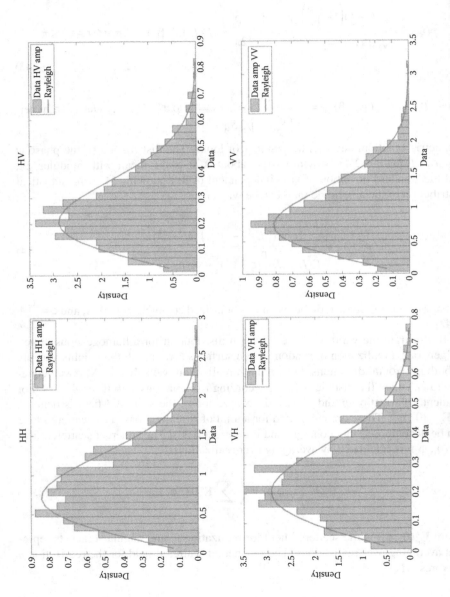

FIGURE 2.9 Comparison of amplitude distributions between simulated data: four polarized channels, HH, VV, HV, and VH, and Rayleigh distribution.

the incoherent wave contributes to radar backscattering. However, for numerical simulations, the surface is not that large and the maximum surface area in this chapter is 32 × 32 wavelengths. The coherent wave has an angular spread, which is of the order of the wavelength divided by L_s. The coherent field is subtracted from the total scattered field to calculate the incoherent field for each realization. Using the above normalization, the average absolute square of the scattering matrix elements will give the backscattering coefficients. Note that in the 3D numerical method of Maxwell's equations (NMM3D), simulations are based on the forward scattering alignment (FSA) convention, of which the incident and scattered wave vectors are opposite those in backscattering, as $\hat{k}_s = -\hat{k}_i$ [9,20]. Figure 2.9 plots the statistical distribution of four polarized simulated data, HH, VV, HV, and VH, along with the Rayleigh distribution. It is confirmed that the simulated data well preserve the speckle properties. From Figure 2.10, the statistical distribution of the phase difference between HH and VV polarizations is shown. It is clear that the match

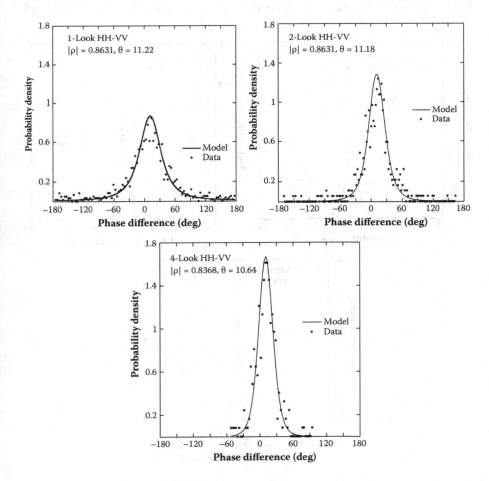

FIGURE 2.10 Comparison of statistical distribution of phase difference of HH and VV polarization for 1, 2, and 4 looks.

between Lee's model prediction (Equation 2.43) and SAR simulated data is excellent for 1, 2, and 4 looks. The decreasing of the dispersion when the number of looks is increasing is shown, as expected. To further confirm the statistical property, Figure 2.11 displays the phase difference distributions between HV and VH polarizations. Again, an excellent match is observed for 1, 2, and 4 looks. It is worth noting that simulated data for cross-polarization are accurate even though their level of amplitude is relatively low compared to the co-polarization response. Next, we investigate the distribution of the normalized amplitude ratio between HH and VV polarizations. Figure 2.12 compares the distribution of the amplitude ratio of HH and VV polarized channels between the model prediction and simulated data. It is shown that the model and data are in good agreement. Similarly, the amplitude ratio distribution of HV and VH is plotted in Figure 2.13. Again, excellent agreement between simulated data and the model is obtained. This suggests that the SAR simulations developed in this chapter are fully capable of simulating the fully polarimetric responses from randomly rough surfaces, and thus offer a viable tool

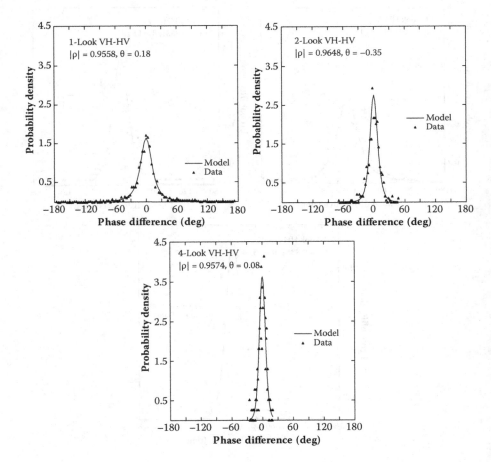

FIGURE 2.11 Comparison of statistical distribution of phase difference of HV and VH polarization for 1, 2, and 4 looks.

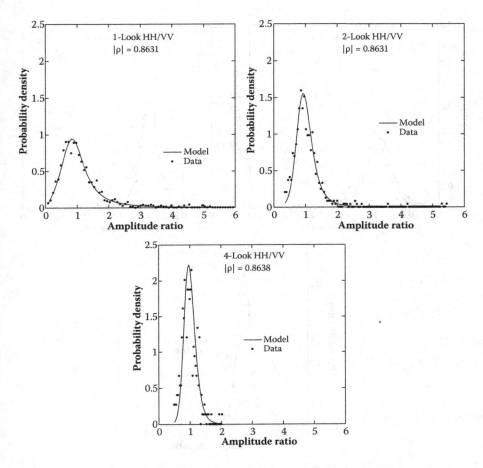

FIGURE 2.12 Statistical distribution of amplitude ratio between HH and VV polarizations for 1, 2, and 4 looks.

to study the characteristics of polarimetric descriptors of rough surface scattering, to be presented in what follows.

2.3 TARGET RCS MODELS

2.3.1 DETERMINISTIC TARGET

To account for the radar response from a deterministic target, we need the target's radar cross section under the radar observation scenario. The coherent scattering process between the target and its background is neglected for the sake of simplicity. In [23], the physical diffraction theory (PDT) [43] and shooting and bouncing rays (SBRs) were implemented to compute the RCS of complex radar targets [11,12,40,41, 44]. It is noted that SBR implements the computation of multilevel electric field (EF) reflection and refraction in the electric field domain; physical optics provides

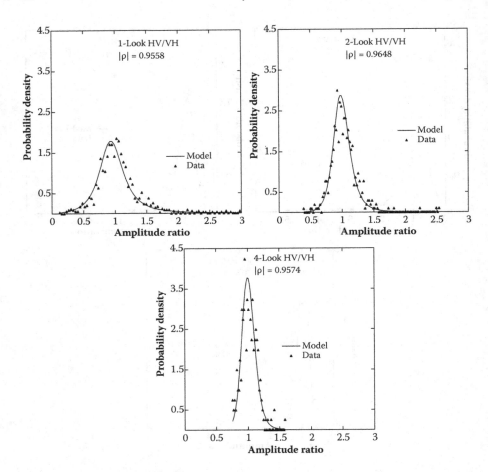

FIGURE 2.13 Statistical distribution of amplitude ratio between HV and VH polarizations for 1, 2, and 4 looks.

a first-order approximate tangential surface field induced by the incident field, and geometric optics deals with the direction and energy propagation of the electric field. More rigorous frequency-domain (e.g., moment method–based) and time-domain (e.g., finite difference time domain [FDTD]-based) algorithms may be applied but require much more computational resources. Commercial software packages are available, such as FEKO, HFSS, and ADS.

For a given SAR observation geometry, an incident plane is formed from which an electromagnetic wave impinges upon the targets. In ray tracing, we need to determine whether there are object intersections in the target CAD model. To follow this, the bounding volume hierarchy (BVH) [45–47] is adopted to remove or eliminate potential intersections within the bounding volume. In BVH, a tree structure on a set of geometric CAD models, the number of levels is determined by the incident wavelength, as discussed in Section 2.2.1. The target CAD model contains numerous grids or polygons, each associated with computed RCS as a function of incident and aspect angles for a

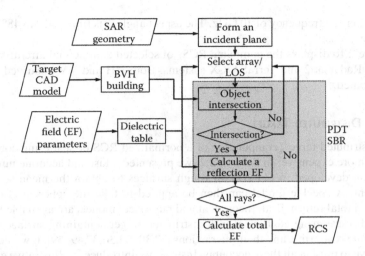

FIGURE 2.14 Block diagram of RCS computation using PDT and SBR for a given SAR geometry and electric field parameters.

given set of radar parameters. For each polygon, the dielectric constant of the target material corresponding to the electric field parameters is obtained through a lookup dielectric table. The number of polygons is determined by the target's geometry complexity and electromagnetic size. To realize the imaging scenario, each polygon must be properly oriented and positioned based on earth centered rotational (ECR) coordinates, as will be discussed and treated in Chapter 4. Figure 2.14 is a flowchart of RCS computation as outlined above. Noted that other numerical electromagnetic simulation schemes may replace the steps of SBR and PDT.

Figure 2.15 provides the RCS computation of a dihedral corner reflector to validate the numerical algorithm in Figure 2.14. Reference data are taken from [48] for

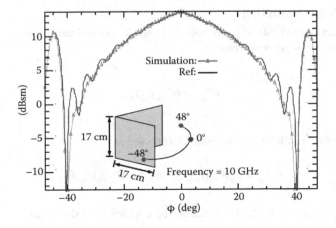

FIGURE 2.15 RCS of dihedral corner reflector as a function of the aspect angle at 10 GHz.

comparison at a frequency of 10 GHz. The aspect angle ϕ is from $-48°$ to $48°$ rotated about the z-axis.

Figure 2.16 displays the computed RCSs of selected commercial aircrafts for the cases of Radarsat-2 and TerraSAR-X systems. Both HH and VV polarized returns are computed.

2.3.2 DISTRIBUTED TARGET

For a distributed target, computation of a normalized RCS or scattering coefficient is much more expensive. In Section 2.2, we presented a fast and accurate numerical simulation developed in [34–37] for rough surfaces to obtain the mean scattering coefficient. A speckle model can then be applied to take the coherent effect into account in total return. Both theoretical and numerical models are available for estimating the scattering coefficient of a distributed target containing surface scattering, volume scattering, and their interactions [17,30–33,36,37,49–59]. It would be too exhaustive to present all these accounts. Instead, we introduce an illustrative example of radar scattering from a vegetation canopy over a rough surface. Reduction to pure surface scattering or pure volume scattering, if inclined, is possible and straightforward. Depending on the spatial resolution, the so-estimated scattering coefficient, after speckle is included, can be put into the SAR system model to simulate the SAR signal, as will be developed in Section 2.4, only to serve as an illustrative example.

Mathematically, we may express the total scattering return from a vegetation canopy as two terms:

$$\sigma^o_{qp} = \sigma^o_{s,qp} + \sigma^o_{d,qp} \tag{2.47}$$

where $\sigma^o_{s,qp}$ represents the single scattering and $\sigma^o_{d,qp}$ the double scattering, ignoring the higher-order scattering [30,31,52]. Figure 2.17 explains the scattering components of the vegetation canopy consisting of single and double scattering.

2.3.2.1 Single Scattering

Single scattering consists of contributions from the ground surface beneath the rice plant and from a vegetative volume:

$$\sigma^o_{s,qp} = \sigma^s_{s,qp} + \sigma^v_{s,qp} \tag{2.48}$$

with the surface scattering contribution expressed as

$$\sigma^s_{s,qp} = L_q(\theta_s)\sigma^o_{g,qp}(\theta_s,\phi_s;\pi-\theta_i,\phi_i)L_p(\theta_i) \tag{2.49}$$

where $\sigma^o_{g,qp}(\theta_s,\phi_s;\pi-\theta_i,\phi_i)$ is the scattering coefficient of the ground surface.

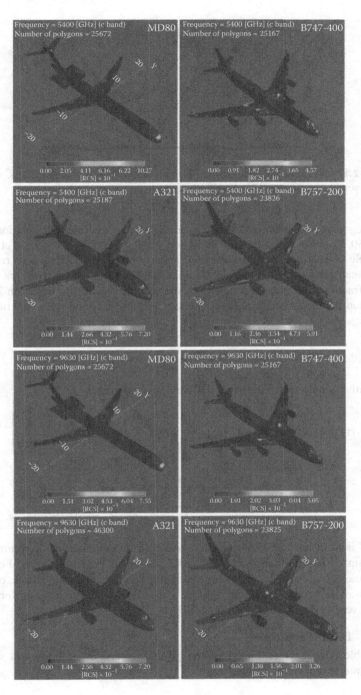

FIGURE 2.16 (See color insert.) Computed RCS of commercial aircrafts for C-band HH Polarization (left) and X-band VV Polarization (right).

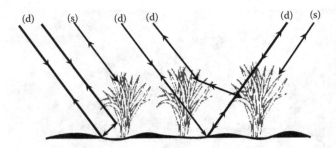

FIGURE 2.17 Illustration of the scattering components from vegetation plants, where (s) indicates single scattering and (d) double scattering.

For scattering from an underlying ground surface, the integral equation model (IEM) [30,31] or advanced integral equation model (AIEM) [49] is a good choice for the simple calculations and produces sufficiently accurate estimates. To radar, the surface scattering contribution may be attenuated, depending on the plant density, volume fraction, and radar wavelength of the vegetative plant. In Equation 2.49, the attenuation factor L_m is determined by the plant height H and the extinction cross section κ_m.

$$L_m(\theta_u) = \exp[-\kappa_m(\theta_u)H \sec \theta_u], \quad m = p, q; u = i, s \tag{2.50}$$

The other single scattering from the vegetative volume is

$$\sigma^v_{s,qp} = 4\pi L_p(\theta_i)Q_{pq}(\theta_s,\phi_s;\pi-\theta_i,\phi_i)\frac{1-L_p(\theta_i)L_q(\theta_s)}{\kappa_p(\theta_i)\sec\theta_i + \kappa_q(\theta_s)\sec\theta_s} \tag{2.51}$$

$$Q_{qp} = \frac{1}{4\pi}\sum_{j=1}^{N}P(n_j)\langle\sigma_j\rangle \tag{2.52}$$

where N is the total number of scatterers within the antenna beam volume, $P(n_j)$ is the number density, and σ_j is radar cross section of the jth scatterer. For a rice canopy, dielectric cylinders are used to model the stems, while the elliptic discs are for leaves. More complex geometry objects may be adopted to account for a more complex vegetation structure [50,53,56,57].

2.3.2.2 Double Scattering

The double bounce is generated by the ground–volume interaction and volume–volume interaction and is expressed as

$$\sigma^o_{d,pq} = \sigma^o_{d,pq}(v \leftrightarrow g) + \sigma^o_{d,pq}(v \leftrightarrow v) \tag{2.53}$$

For co-polarization and lower frequency, the first term is important, while the second term is more important for cross-polarization and higher frequency. The co-polarized interaction is

$$\sigma^o_{d,qq}(v \leftrightarrow g) = 8\pi H L_p^2(\theta_i)\sigma^o_{g,qp}(\theta_s,\phi_s;\pi - \theta_i,\phi_i)Q_{pp}(\theta_i,\phi_i + \pi;\theta_i,\phi_i) \quad (2.54)$$

while for cross-polarization the interaction is

$$\sigma^o_{d,qp}(v \leftrightarrow g) = 8\pi L_q(\theta_i)L_p(\theta_i)\sigma^o_{g,qp}(\theta_s,\phi_s;\pi - \theta_i,\phi_i)Q_{qp}(\theta_i,\phi_i + \pi;\theta_i,\phi_i)$$

$$\frac{L_q(\theta_i) - L_p(\theta_i)}{\kappa_q(\theta_i) - \kappa_p(\theta_i)} \quad (2.55)$$

The scattering process of the volume to volume is much more complex. For more details, the variables and parameters in Equations 2.49 through 2.55 can be found in [30,52].

Figure 2.18 shows C-band polarized backscattering coefficients of a typical rice plant as a function of growth phases: vegetative, reproductive, and ripening, spanning about 120 days.

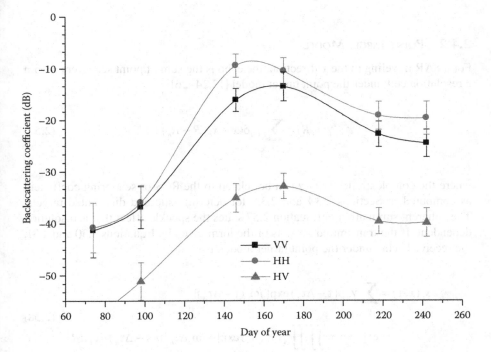

FIGURE 2.18 C-band backscattering coefficient of rice plant as a function of growth phase.

2.4 SYSTEM MODEL

2.4.1 SLANT RANGE PROJECTION

We elaborate more on the SAR operation from a systems point of view. The concept of SAR was briefly introduced in Chapter 1. Though SAR imaging can be made in Stripmap, Spotlight, and ScanSAR modes, and others, we will focus on the side-looking Stripmap mode for its popular uses. As mentioned earlier, SAR measures the complex scattered field $\mathbf{E}(\bar{r}')$ (Figure 2.1). For SAR observation located at (x, y_0, z_0) with look angle θ, the two-dimensional projection of the scattered field is

$$E(x, R) = \int E(x, y_0 + R\sin\theta, z_0 - R\cos\theta)R\, d\theta \tag{2.56}$$

This projection from ground range to slant range introduces distortions of the target reflectivity when we project the received scattered field back to the ground plane. The extent of distortion is strongly dependent on the terrain geometry and radar observation geometry, aside from focusing issues. Commonly encountered and pronounced geometric distortions include shadow, layover, and foreshortening. Care must be exercised when applying the SAR imagery data to interpret and invert the physical properties of the targets.

2.4.2 POINT TARGET MODEL

For a SAR traveling in the x-direction, the echo is the sum of point scatterers within a resolution cell under the point target model [13,24–26]

$$\gamma(x, R) = \sum_n \gamma_n \delta(x - x_n, R - R_n) \tag{2.57}$$

where the complex reflectivity γ can be related to the RCS or scattering coefficient as computed in Sections 2.3.1 and 2.3.2 for deterministic and distributed targets. The coherent summation in Equation 2.57 states the speckle effect that is resolution dependent. If the transmitted pulse is of the form given by Equations 1.30 and 1.31, the received echo, under the point target model, is

$$s_r(x, \tau) = \sum_n \gamma_n p(\tau - \Delta\tau_n) \exp[j\omega_c(\tau - \Delta\tau_n)]$$

$$= \exp(j\omega_c\tau) \iiint \gamma(x_n, R_n) \exp[-j\omega_c\Delta\tau_n] p(\tau - \Delta\tau_n)\, dx_n\, dR_n \tag{2.58}$$

Ignoring the platform motion-induced drifts, the time delay between transmission and reception from the nth target is

$$\Delta \tau_n = \Delta \tau_n(x, x_n, R_n) = \frac{2R(x - x_n, R_n)}{c} \tag{2.59}$$

which obviously is range dependent. The received signal is correlated with a reference signal, a matched filtering process, resulting in [24,25]

$$v(x, R_n) = s_r^n(x, x_n, R_n) \otimes_x s_{ref}(x, R_n)$$

$$= \int_{R_n - L_{sa}/2}^{R_n + L_{sa}/2} \gamma_n \exp\left[\frac{2j\omega_c R_n}{c}\right] \exp\left[\frac{j\omega_c(\xi - x_n)^2}{cR_n}\right] \exp\left[-\frac{j\omega_c(\xi - x)^2}{cR_n}\right] d\xi$$

$$= \gamma_n L_{sa} \exp\left[\frac{2j\omega_c R_n}{c}\right] \exp\left[\frac{j\omega_c(x - x_n)^2}{cR_n}\right] \text{sinc}\left[\frac{\omega_c L_{sa}}{cR_n}(x - x_n)\right] \tag{2.60}$$

where L_{sa} is the synthetic aperture length (see Figure 1.5).

In the Equation 2.60, the first and third terms on the right-hand side represent the coupling of the range and azimuth and are separable: $R(x - x_n, R_n) = \sqrt{R_n^2 + (x - x_n)^2}$. The limited observation time (see Figure 1.5) is contained in $\text{sinc}\left[\frac{\omega_c L_{sa}}{cR_n}(x - x_n)\right]$. The azimuth resolution, after azimuth processing, is

$$\delta_{az} = \frac{\pi c R_n}{\omega_c L_{sa}} = \frac{\lambda R_n}{2L_{sa}} = \frac{\ell_{ra}}{2} \tag{2.61}$$

which is identical to Equation 1.59. When the observation time approaches infinity, $t_{az} \to \infty$, that is, $L_{sa} \to \infty$, it is found that

$$\text{sinc}\left[\frac{\omega_c L_{sa}}{cR_n}(x - x_n)\right] \to \delta(x - x_n) \tag{2.62}$$

That is, the system response is the delta function, an ideal but unrealistic case. The output signal shows a strong coupling between the range and azimuth. Also note that the phase in the first term is only useful in interferometric processing.

Another way to consider the SAR from a systems point of view is using Equation 1.2. The two-dimensional (azimuth–range) SAR impulse response function is recognized as

$$h_{SAR}(x, R) = \text{sinc}(x/\Delta x)\text{sinc}(R/\Delta R) \tag{2.63}$$

where ΔR, Δx, as given in Equation 2.63 are the resolutions in the range and azimuth directions, respectively. When the resolution approaches zero (ideal), $h_{SAR}(x, R)$ becomes a delta function, which perfectly reconstructs the target reflectivity. Also, in defining the above function, we ignore the range migration effect such that the range and slow time are linearly related. Of course, this assumption is only for easy discussion to grasp the SAR operation concept. By matched filter, the echo signal becomes

$$s(x, R) = \gamma(x, R)\exp(-2jkR) \otimes_x \otimes_R h_{SAR}(x, R) + n(x, R)$$
$$= \iint \gamma(x', R')e^{-2jkR'} h_{SAR}(x - x', R - R')\,dx'\,dR' + n(x, R) \tag{2.64}$$

where $n(x, R)$ is the system noise.

The slant range in Equation 2.56 is a function of slow time η,

$$R_n(\eta) = R_0 + \frac{\partial R_0}{\partial \eta}(\eta - \eta_0) + \frac{\partial^2 R_0}{\partial \eta^2}(\eta - \eta_0)^2/2 + \dots \tag{2.65}$$

where the second term on the right-hand side is related to Doppler frequency, while the third term is related to the Doppler rate. The first-order time dependence of the range is called the range walk, and the second-order time dependence is the range curvature. The sum of the range walk and range curvature is the range cell migration. Correction of range cell migration is one of the major tasks in image focusing and will be thoroughly treated in Chapter 6. Notice that in some extreme cases (highly maneuverable), the expansion of terms in Equation 2.65 to higher than the second order may be necessary. Recall that the phase associated with the range in Equation 2.65 has been discussed in Chapter 1.

APPENDIX: DERIVATION OF MULTI-LOOK AMPLITUDE DISTRIBUTION

For easy reference, this appendix gives derivations of Equations 2.35 and 2.37, though they might be found from some reference books [9,10,27]. From Equation 2.34, let

$$z = LI = \sum_{k=1}^{L} I_k \tag{A.1}$$

Then, the probability density function of z is [60]

$$p_z = p_{I_1} \otimes p_{I_2} \otimes p_{I_3} \cdots \otimes p_{I_L} \tag{A.2}$$

where $\otimes$ is the convolution operator and $p_{I_i}, i = 1, 2, ..L$, is the probability function of the ith look intensity. By means of a characteristic function, Equation A.2 is mathematically equivalent to the expression

$$\Phi_z(\omega) = \Phi_{I_1}(\omega) \times \Phi_{I_2}(\omega) \times \Phi_{I_3}(\omega) \cdots \times \Phi_{I_L}(\omega) = [\Phi_I(\omega)]^L \quad \text{(A.3)}$$

where

$$\Phi_I(\omega) = \int_{-\infty}^{\infty} e^{i\omega I} p_I(I) \, dI \equiv \langle e^{i\omega I} \rangle \quad \text{(A.4)}$$

By substituting Equation 2.29 into Equation A.4, we have

$$\Phi_I(\omega) = \int_{-\infty}^{\infty} e^{i\omega I} p_I(I) \, dI = \int_{-\infty}^{\infty} e^{i\omega I} \frac{1}{\sigma^2} \exp\left(-\frac{I}{\sigma^2}\right) dI = \frac{1/\sigma^2}{1/\sigma^2 - j\omega}, \quad I \geq 0 \quad \text{(A.5)}$$

It follows that

$$[\Phi_I(\omega)]^L = \sigma^{-2L}[1/\sigma^2 - j\omega]^{-L} \quad \text{(A.6)}$$

To find p_z, we take the inverse Fourier transform of Equation A.6, reaching to [61]

$$p_z(z) = \frac{1}{2\pi} \int_{-\infty}^{+\infty} \Phi_z(\omega) e^{-j\omega z} \, d\omega = \frac{1}{2\pi} \sigma^{-2L} \int_{-\infty}^{+\infty} [1/\sigma^2 - j\omega]^{-L} e^{-j\omega z} \, d\omega \quad \text{(A.7)}$$

Using the identity [38,39]

$$\int_{-\infty}^{+\infty} [\beta - jx]^{-v} e^{-jpx} \, dx = \frac{2\pi p^{v-1}}{\Gamma(v)} e^{-\beta p} \quad \text{(A.8)}$$

Equation A.7 becomes

$$p_z(z) = \sigma^{-2L} \frac{z^{L-1}}{\Gamma(L)} e^{-z/\sigma^2} \quad \text{(A.9)}$$

Finally, we obtain Equation 2.35 by changing the variable back:

$$p_I(I) = L p_z(z = LI) = \left(\frac{L}{\sigma^2}\right)^L \frac{1}{\Gamma(L)} \exp\left(-\frac{LI}{\sigma^2}\right) I^{L-1} \quad \text{(A.10)}$$

For amplitude distribution, notice that

$$p_A(A)dA = p_I(I)\big|_{I=A^2} \, dI \tag{A.11}$$

With $dA = \dfrac{1}{2} I^{-1/2} dI$, Equation A.11 becomes

$$p_A(A \mid \sigma^2) = 2\left(\frac{L}{\sigma^2}\right)^L \frac{A^{(2L-1)}}{\Gamma(L)} \exp\left(-\frac{LA^2}{\sigma^2}\right), \quad A \geq 0 \tag{A.12}$$

which is Equation 2.37.

REFERENCES

1. Barton, D. K., *Modern Radar System Analysis*, Artech House, Norwood, MA, 1988.
2. Berkowitz, R. S., ed., *Modern Radar: Analysis, Evaluation, and System Design*, John Wiley & Sons, New York, 1965.
3. Blackledge, J. M., *Quantative Coherent Imaging: Theory, Imaging, and Some Applications*, Academic Press, New York, 1989.
4. Chen, C. H., ed., *Information Processing for Remote Sensing*, World Scientific Publishing Co., Singapore, 2000.
5. Chen, C. H., ed., *Signal and Image Processing for Remote Sensing*, 2nd ed., CRC Press, New York, 2012.
6. Cloude, S. R., *Polarisation: Applications in Remote Sensing*, Oxford University Press, Oxford, 2009.
7. Elachi, C., *Space-Borne Radar Remote Sensing: Applications and Techniques*, IEEE Press, New York, 1988.
8. Jin, Y. Q., and Xu, F., *Polarimetric Scattering and SAR Information Retrieval*, Wiley-IEEE Press, New York, 2013.
9. Lee, J.-S., and Pottier, E., *Polarimetric Radar Imaging: From Basics to Applications*, CRC Press, New York, 2009.
10. Oliver, C., and Quegan, S., *Understanding Synthetic Aperture Radar Images*, SciTech Publishing, Raleigh, NC, 2004.
11. Rihaczek, A. W., and Hershkowitz, S. J., *Radar Resolution and Complex-Image Analysis*, Artech House, Norwood, MA, 1996.
12. Shirman, Y. D., ed., *Computer Simulation of Aerial Target Radar Scattering, Recognition, Detection, and Tracking*, Artech House, Norwood, MA, 2002.
13. Skolnik, M. I., *Introduction to Radar Systems*, McGraw-Hill, Singapore, 1981.
14. Skolinik, M. I., ed., *Radar Handbook*, 3rd ed., McGraw-Hill, New York, 2008.
15. Soumekh, M., *Synthetic Aperture Radar Processing*, John Wiley & Sons, New York, 1999.
16. Sullivan, R. J., *Microwave Radar: Imaging and Advanced Concepts*, Artech House, Norwood, MA, 2000.
17. Ulaby, F. T., Moore, R. K., and Fung, A. K., *Microwave Remote Sensing: Active and Passive*, vol. 2, *Radar Remote Sensing and Surface Scattering and Emission Theory*, Artech House, Norwood, MA, 1982.
18. Ulaby, F. T., Moore, R. K., and Fung, A. K., *Microwave Remote Sensing: Active and Passive*, vol. 3, Artech House, Norwood, MA, 1986.

19. Ulaby, F. T., and Elachi, C., eds., *Radar Polarimetry for Geosceience Applications*, Artech House, Norwood, MA, 1990.
20. Ulaby, F. T., and Long, D. G., *Microwave Radar and Radiometric Remote Sensing*, University of Michigan Press, Ann Arbor, 2013.
21. van Zyl, J. J., *Synthetic Aperture Radar Polarimetry*, Wiley, New York, 2011.
22. Cook, C. E., *Radar Signals: An Introduction to Theory and Applications*, Artech House, Norwood, MA, 1993.
23. Chen, K. S., and Tzeng, Y. C., On SAR image processing: From focusing to target recognition, in *Signal and Image Processing for Remote Sensing*, ed. C. H. Chen, 2nd ed., CRC Press, New York, 2012.
24. Cumming, I., and Wong, F., *Digital Signal Processing of Synthetic Aperture Radar Data: Algorithms and Implementation*, Artech House, Norwood, MA, 2004.
25. Curlander, J. C., and McDonough, R. N., *Synthetic Aperture Radar: Systems and Signal Processing*, Wiley-Interscience, New York, 1991.
26. Franceschetti, G., and Lanari, R., *Synthetic Aperture Radar Processing*, CRC Press, Boca Raton, FL, 1999.
27. Maitre, H., ed., *Processing of Synthetic Aperture Radar (SAR) Images*, Wiley-ISTE, New York, 2008.
28. Margarit, G., Mallorqui, J. J., Rius, J. M., and Sanz-Marcos, J., On the usage of GRECOSAR, an orbital polarimetric SAR simulator of complex targets, to vessel classification studies, *IEEE Transactions on Geosciences and Remote Sens*ing, 44: 3517–3526, 2006.
29. Wehner, D. R., *High-Resolution Radar*, Artech House, Norwood, MA, 1995.
30. Fung, A. K., *Microwave Scattering and Emission Models and Their Applications*, Artech House, Norwood, MA, 1994.
31. Fung, A. K., and Chen, K. S., *Microwave Scattering and Emission Models for Users*, Artech House, Norwood, MA, 2011.
32. Tsang, L., Kong, J. A., and Shin, R. T., *Theory of Microwave Remote Sensing*, Wiley-Interscience, New York, 1985.
33. Chen, K. S., Tsang, L., Chen, K. L., Liao, T. H., and Lee, J. S., Polarimetric simulations of SAR at L-band over bare soil using scattering matrices of random rough surfaces from numerical three-dimensional solutions of Maxwell's equations, *IEEE Transactions on Geosciences and Remote Sensing*, 52(11): 7048–7058, 2014.
34. Huang, S., Tsang, L., Njoku, E. G., and Chen, K. S., Backscattering coefficients, coherent reflectivities, emissivities of randomly rough soil surfaces at L-band for SMAP applications based on numerical solutions of Maxwell equations in three-dimensional simulations, *IEEE Transactions on Geosciences and Remote Sens*ing, 48: 2557–2567, 2010.
35. Huang, S., and Tsang, L., Electromagnetic scattering of randomly rough soil surfaces based on numerical solutions of Maxwell equations in 3 dimensional simulations using hybrid UV/PBTG/SMCG method, *IEEE Transactions on Geosciences and Remote Sensing*, 50: 4025–4035, 2012.
36. Tsang, L., Kong, J. A., Ding, K. H., and Ao, C. O., *Scattering of Electromagnetic Waves*, vol. 2, *Numerical Simulations*, Wiley Interscience, Hoboken, NJ, 2001.
37. Tsang, L., Ding, K. H., Huang, S. H., and Xu, X., Electromagnetic computation in scattering of electromagnetic waves by random rough surface and dense media in microwave remote sensing of land surfaces, *Proceedings of IEEE*, 101: 255–279, 2013.
38. Abramowitz, M., and Stegun, I. A., *Handbook of Mathematical Functions: With Formulas, Graphs, and Mathematical Tables*, Dover, New York, 1972.
39. Gradshteyn, I., and Ryzhik, I., *Table of Integrals, Series and Products*, Academic Press, New York, 2000
40. Lee, S. W., Ling, H., and Chou, R., Ray-tube integration in shooting and bouncing ray method, *Microwave and Optical Technology Letters*, 1: 286–289, 1988.

41. Lin, H., Chou, R. C., and Lee, S. W., Shooting and bouncing rays: Calculating the RCS of an arbitrarily shaped cavity, *IEEE Transactions on Antennas and Propagation*, 37: 194–205, 1989.
42. Lee, J.-S., Hoppel, K. W., Mango, S. A., and Miller, A. R., Intensity and phase statistics of multi-look polarimetric and interferometric SAR imagery, *IEEE Transactions on Geosciences and Remote Sensing*, 32(5): 1017–1028, 1994.
43. Jeng, S. K., Near-field scattering by physical theory of diffraction and shooting and bouncing rays, *IEEE Transactions on Antennas and Propagation*, 46: 551–558, 1998.
44. Bhalla, R., and Ling, H., 3D scattering center extraction using the shooting and bouncing ray technique, *IEEE Transactions on Antennas and Propagation*, 44: 1445–1453, 1996.
45. Ericson, C., *Real-Time Collision Detection*, CRC Press, New York, 2004.
46. Günther, J., Popov, S., Seidel, H.-P., and Slusallek, P., Realtime ray tracing on GPU with BVH-based packet traversal, *IEEE Symposium on Interactive Ray Tracing*, 113–118, 2007.
47. Hughes, J. F., and van Dam, A., *Computer Graphics: Principles and Practice*, 3rd ed., Addison-Wesley, New York, 2013.
48. Peixoto, G. G., De Paula, A. L., Andrade, L. A., Lopes, C. M. A., and Rezende, M. C., Radar absorbing material (RAM) and shaping on radar cross section reduction of dihedral corners, *SBMO/IEEE MTT-S International Conference on Microwave and Optoelectronics*, 460–463, 2005.
49. Chen, K. S., Wu, T. D., Tsang, L., Li, Q., Shi, J. C., and Fung, A. K., Emission of rough surfaces calculated by the integral equation method with comparison to three-dimensional moment method simulations, *IEEE Transactions on Geosciences and Remote Sensing*, 41(1): 90–101, 2003.
50. Du, Y., Luo, Y., Yan, W.-Z., and Kong, J. A., An electromagnetic scattering model for soybean canopy, *Progress in Electromagnetics Research*, 79: 209–223, 2008.
51. Fung, A. K., Li, Z. Q., and Chen, K. S., Backscattering from a randomly rough dielectric surface, *IEEE Transactions on Geosciences and Remote Sensing*, 30: 356–369, 1992.
52. Karam, M. A., Fung, A. K., Lang, R. H., and Chauhan, N. S., A microwave scattering model for layered vegetation, *IEEE Transactions on Geosciences and Remote Sensing*, 30: 767–784, 1992.
53. Koh, I. S., and Sarabandi, K., A new approximate solution for scattering by thin dielectric disks of arbitrary size and shape, *IEEE Transactions on Antennas and Propagation*, 53: 1920–1926, 2005.
54. Sun, G., and Ranson, K. J., A three-dimensional radar backscatter model of forest canopies, *IEEE Transactions on Geoscience and Remote Sensing*, 33: 372–382, 1995.
55. Ulaby, F. T., Sarabandi, K., McDonald, K., Whitt, M., and Dobson, M. C., Michigan microwave canopy scattering model, *International Journal of Remote Sensing*, 11(7): 1223–1253, 1990.
56. Yan, W.-Z., Du, Y., Wu, H., Liu, D., and Wu, B.-I., EM scattering from a long dielectric circular cylinder, *Progress in Electromagnetics Research*, 85: 39–67, 2008.
57. Yan, W.-Z., Du, Y., Li, Z., Chen, E.-X., and Shi, J.-C., Characterization of the validity region of the extended T-matrix method for scattering from dielectric cylinders with finite length, *Progress in Electromagnetics Research*, 96: 309–328, 2009.
58. Yueh, S. H., Kong, J. A., Jao, J. K., Shin, R. T., and Le Toan, T., Branching model for vegetation, *IEEE Transactions on Geoscience and Remote Sensing*, 30: 390–401, 1992.
59. Zhang, R., Hong, J., and Ming, F., SAR echo and image simulation of complex targets based on electromagnetic scattering, *Journal of Electronics and Information Technology*, 32(12): 2836–2841, 2010.
60. Brown, R., and Hwang, P., *Introduction to Random Signal Analysis and Kalman Filtering*, Wiley, New York, 1983.
61. Strang, G., *Introduction to Applied Mathemathics*, Cambridge Press, Wellesley, MA, 1986.

3 SAR Data and Signal

3.1 INTRODUCTION

The signal property of synthetic aperture radar (SAR) is one of the core issues to grasp in order to process the raw data into an image. The mapping between the object domain and image domain is persistently bridged and thus controlled by the signal domain. This chapter deals with this aspect in depth. Although in SA, various waveforms, like continuous wave (CW) and phase-coded [1–3], may be used, we will focus our discussion on linear frequency modulation (LFM) (chirp) and frequency-modulated continuous wave (FMCW) for their common use in modern SAR systems.

3.2 CHIRP SIGNAL

3.2.1 Echo Signal in Two Dimensions

In Equation 1.54, we indicated that the pulse repetition frequency f_p must follow the Nyquist criteria to avoid the range ambiguity within the maximum probing range R_{max}. Also, there is a minimum range R_{min} corresponding to the near range. The antenna elevation beamwidth β_e subtends from the near to the far range (swath) (Figure 3.1). To avoid the air bubbles, SAR is not pointing its boresight at the nadir. The delay time of returns from the near range and far range, τ_{near}, τ_{far}, must be confined within a period. Mathematically, the following two relations should be obeyed [4,5]:

$$T_p < \frac{2R_{min}}{c} \tag{3.1}$$

$$T = \frac{1}{f_p} > \frac{2R_{max}}{c} + T_p \tag{3.2}$$

Remember that for spaceborne SAR, the maximum range is limited by the earth curvature [6,7], depending on the satellite height and elevation angle. If the transmitted signal is of the form given by Equation 1.54, the received signal as a function of fast time is the output of the matched filter:

$$s_r(\tau) = A_0 \otimes s_t(\tau) = \int_{t-T_p/2}^{t+T_p/2} A_0(t) s_t(\tau - t)\, dt$$

$$= A_0 p_r \left(\tau - \frac{2R}{c} \right) \cos \left\{ 2\pi f_c \left(\tau - \frac{2R}{c} \right) + \pi a_r \left(\tau - \frac{2R}{c} \right)^2 \right\} \tag{3.3}$$

51

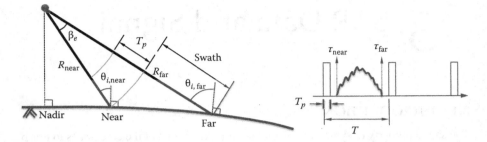

FIGURE 3.1 Observation geometry defined by pulse and antenna beamwidth.

where A_0 is the scatter amplitude [4] and, without loss of generality, is assumed constant. Equation 3.3 is indeed also dependent on the slow time because the slant range R is varying with the sensor position within the target exposure time. Referring to Figure 1.6, the slow time-dependent slant range can be expanded about $R(\eta_c)$, with the beam center crossing time, η_c:

$$R(\eta) = R(\eta_c) + \frac{u^2 \eta_c}{R(\eta_c)}(\eta - \eta_c) + \frac{1}{2}\frac{u^2 \cos^2 \theta_{\ell,c}}{R(\eta_c)}(\eta - \eta_c)^2 + \dots \qquad (3.4)$$

where θ_c is look angle to the scene center and

$$\eta_c = \frac{R_0 \tan \theta_c}{u} = \frac{R(\eta_c)\sin \theta_c}{u}. \qquad (3.5)$$

Due to the time variation of the range, a point target response will continuously appear along the path according to Equation 3.4 within the synthetic aperture length. A coherent sum of these responses during the course of the target exposure time will be out of focus if no range-induced phase variation is corrected. Figure 3.2

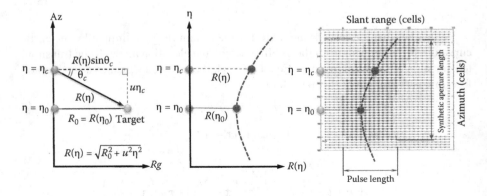

FIGURE 3.2 Mapping from the object domain to the data domain.

schematically illustrates data collection and mapping from the target (object) to the SAR data domain.

In practice, we should take the antenna pattern into the signal reception. The received signal may be written as

$$s_r(\tau,\eta) = A_0 p_r\left(\tau - \frac{2R}{c}, T_p\right) g_a(\eta)\cos\left\{2\pi f_c\left(\tau - \frac{2R(\eta)}{c}\right) + \pi a_r\left(\tau - \frac{2R(\eta)}{c}\right)^2\right\}$$

(3.6)

A typical two-way antenna pattern in the azimuthal direction may be of the form [4]

$$g_a(\eta) \cong \mathrm{sinc}^2\left\{\frac{0.886\theta_{\mathrm{diff}}(\eta)}{\beta_{az}}\right\}$$

(3.7)

where $\theta_{\mathrm{diff}} = \theta_{sq} - \theta_{sq,c}$, with $\theta_{sq,c} = \theta_{sq}(\eta_c)$ being the squint angle at the scene center. For small squint approximation, Equation 3.7 may be written as

$$g_a(\eta) = \mathrm{sinc}^2\left(\frac{0.886[\theta_{sq}(\eta) - \theta_{sq,c}]}{\beta_{az}}\right) \cong \mathrm{sinc}^2\left(\frac{0.886}{\beta_{az}}\tan^{-1}\left[\frac{u(\eta - \eta_c)}{R_0}\right]\right).$$

(3.8)

Figure 3.3 plots a right side-looking SAR (left) to show the antenna pattern illuminating on a point target with gain variations (middle) along the azimuth direction; on the right is the intensity of the point target within the cell. The antenna gain pattern effect along the azimuthal direction is clearly observed.

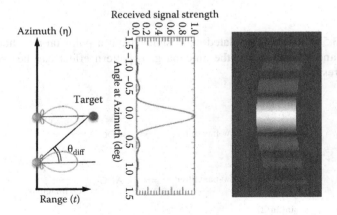

FIGURE 3.3 A right side-looking SAR (left) to show antenna beam pattern (middle) on a point target response (right).

3.2.2 DEMODULATED ECHO SIGNAL

The demodulation is to remove the carrier frequency, which bears no target information. Following [4], a typical demodulation scheme is illustrated in Figure 3.4, where the echo signal in ① is downconverted by mixing with a reference signal, resulting in ② and ③. By filtering out the upper frequency component and going through an analog-to-digital conversion (ADC) if necessary, the echo signal for postprocessing ⑥ is obtained.

①: $\cos\left(2\pi f_c\tau - \dfrac{4\pi f_c R(\eta)}{c} + \pi a_r\left(\tau - \dfrac{2R(\eta)}{c}\right)^2\right) = \cos[2\pi f_c\tau + \phi(\tau)]$

②: $\dfrac{1}{2}\cos[\phi(\tau)] + \dfrac{1}{2}\cos[4\pi f_c\tau + \phi(\tau)]$

③: $\dfrac{1}{2}\sin[\phi(\tau)] + \dfrac{1}{2}\sin[4\pi f_c t + \phi(\tau)]$

④: $\dfrac{1}{2}\cos[\phi(\tau)]$

⑤: $\dfrac{1}{2}\sin[\phi(\tau)]$

⑥: $\dfrac{1}{2}\exp\{j\phi(\tau)\} = \dfrac{1}{2}\exp\left\{-j\dfrac{4\pi f_c R(\eta)}{c} + j\pi a_r\left[t - \dfrac{2R(\eta)}{c}\right]^2\right\}$

The final demodulated two-dimensional echo signal is of the form

$$s_0(\tau,\eta) = A_0 p_r\left(\tau - \dfrac{2R}{c}, T_r\right) g_a(\eta)\exp\left\{-j\dfrac{4\pi f_c R(\eta)}{c} + j\pi a_r\tau\left(t - \dfrac{2R(\eta)}{c}\right)^2\right\}$$

$$(3.9)$$

Figure 3.5 plots a demodulated echo signal from a point target, showing both amplitude and phase. Again, the antenna gain pattern effect can be seen on the amplitude response.

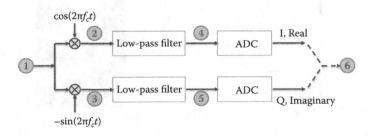

FIGURE 3.4 Demodulation of the echo signal.

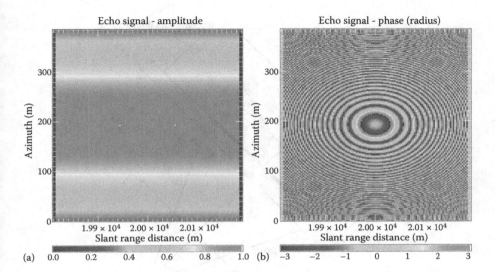

FIGURE 3.5 **(See color insert.)** Typical echo signal after demodulation: amplitude (a) and phase (b).

3.2.3 RANGE CELL MIGRATION

When the imaging taken from the object domain to the data domain is within the synthetic aperture, as illustrated in Figure 3.6, range cell migration (RCM) occurs [4,5,8–10]. This is the energy from a point target response, supposed to be confined at R_0, spreading along the azimuth direction when SAR is traveling. This is easily understood from the slow time-varying range $R(\eta)$, as already illustrated in Figure 3.2. The phase variations associated with slant range, which changes with SAR moving, consist of constant phase, linear phase, quadratic phase, and higher-order terms. The RCM is contributed mainly from the quadratic phase term. This quadratic phase term also determines the depth of focus. For a given synthetic aperture size, the maximum quadratic phase was given in Equations 1.65 and 1.66. Detailed discussion on the phase error and its impact can be found in [4,8,10]. Due to the energy spreading, the effects of the quadratic phase, if not properly compensated, cause image defocusing, lower gain, and higher side lobes in the point target response; that is, in image quality, spatial resolution is degraded with higher fuzziness. Notice that the phase variations, up to the second order, due to the slow-time-dependent range are low-frequency components compared to the phase error in pulse-to-pulse, modulation error and nonlinearity of frequency modulation, which are high frequency since they are induced in the fast-time domain.

The total range migration during the course of synthetic aperture exposure time T_a can be estimated more explicitly from the geometry relation for the cases of low-squint and high-squint angles, as shown in Figure 3.7.

In the low-squint case, the total range of migrations is

$$\Delta R_{\text{total}} = R\left(\eta_c + \frac{T_a}{2}\right) - R(\eta_c) \cong \frac{u^2}{2R_0}\left[\left(\eta_c + \frac{T_a}{2}\right)^2\right] \qquad (3.10)$$

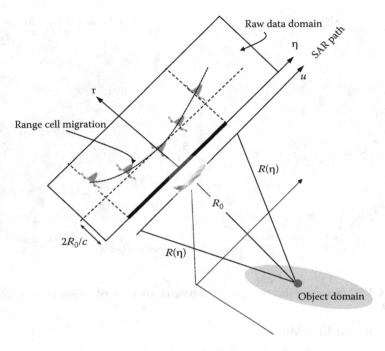

FIGURE 3.6 Illustration of the object domain and data domain showing the range of cell migration.

while for high squint we have

$$\Delta R_{\text{total}} \cong \frac{u^2}{2R_0} T_a \eta_c \tag{3.11}$$

where the exposure time was given previously as

$$T_a = 0.886 \frac{\lambda R(\eta_c)}{L_a u \cos \theta_{sq,c}} \tag{3.12}$$

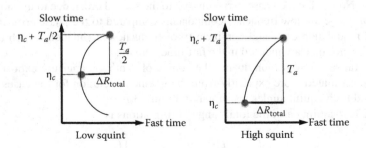

FIGURE 3.7 Total range migrations during the course of synthetic aperture.

The Doppler frequency is obviously also a function of the slow time. Measured at the scene center, the Doppler centroid takes the expression

$$f_{\eta_c} = -\frac{2}{\lambda}\frac{\partial R(\eta)}{\partial \eta}\bigg|_{\eta=\eta_c} = -\frac{2u\eta_c}{\lambda R(\eta_c)} = \frac{2u\sin\theta_{sq,c}}{\lambda} \tag{3.13}$$

and the Doppler rate is

$$a_a = \frac{2}{\lambda}\frac{\partial^2 R(\eta)}{\partial \eta^2}\bigg|_{\eta=\eta_c} = \frac{2u^2\cos^2\theta_{sq,c}}{\lambda R(\eta_c)} \tag{3.14}$$

The Doppler rate as defined is sometimes also called the azimuth FM rate [4] for a stationary target. Note that the Doppler change is only caused by SAR moving along the azimuthal direction. According to Equation 1.61, the total Doppler bandwidth is

$$B_{df} = |a_a| T_a = 0.886\frac{2u\cos\theta_{sq,c}}{\ell_{ra}} \tag{3.15}$$

With the total Doppler bandwidth in mind, it is noted that the pulse repetition frequency (PRF) must be selected to meet the Nyquist criteria [11–13], namely,

$$f_p > B_{df} = 0.886\frac{2u\cos\theta_{sq,c}}{\ell_{ra}} \tag{3.16}$$

Together with Equations 3.1 and 3.2, Equation 3.16 poses constraints on the selection. Under this condition, the selection of PRF must usually be compromised to meet certain requirements [1,2,14,15]. For example, in order to avoid interference from repeated pulse eclipsing, PRF must be such that

$$\frac{(n-1)}{\left(\frac{2}{c}R_{near} - T_p\right)} < \text{PRF} < \frac{n}{\left(\frac{2}{c}R_{far} + T_p\right)} \tag{3.17}$$

where n is the nth pulse, while to avoid the overlapping (indifferentiable) nadir returns, PRF must be chosen by

$$\frac{(n-1)}{\left(\frac{2}{c}R_{near} - T_p - \frac{2}{c}h\right)} < \text{PRF} < \frac{n}{\left(\frac{2}{c}R_{far} + T_p - \frac{2}{c}h\right)} \tag{3.18}$$

where h is the sensor height. Equations 3.17 and 3.18 together implicitly state that there is an available zone of PRF to be selected for a given antenna size, imaging swath, incident angle, and sensor height. Also, bear in mind that the PRF determines the azimuth ambiguity-to-signal ratio for a give Doppler bandwidth.

Other important parameters associated with the two-dimensional raw data domain, such as the range bandwidth, azimuth resolution, and range resolution, can be reexpressed accordingly by taking the antenna gain pattern and squint angle effects into account. Inclined readers are referred to [4,5] for excellent treatment.

3.3 PROPERTIES OF DOPPLER FREQUENCY

3.3.1 SQUINT EFFECTS

For a side-looking SAR system, there always exists a squint angle that is the angle between the antenna boresight and flight path minus 90°. The squint angle induces residual Doppler and causes the shift of the Doppler centroid, as shown in Figure 3.8, where for numerical illustration, we set $f_p = 1000$ kHz, $T = 4 \times 10^{-6}$ s, and $\eta_c = 10^{-6}$ s $= T/4$. Accordingly, the shift of the Doppler centroid is $f_{dc} \doteq f_p/4$ Hz.

3.3.2 FM SIGNAL ALIASING

If the sampling frequency is lower than or equal to the PRF, signal aliasing results [11,12]. This aliasing will cause a "ghost image" after azimuth compression. Beyond $\Delta\eta_{PRF}$, there exist FM signal aliased regions.

$$\Delta\eta_{PRF} = f_p \left| \frac{d\eta}{df_\eta} \right|_{\eta=\eta_c} = \frac{f_p}{a_a} \tag{3.19}$$

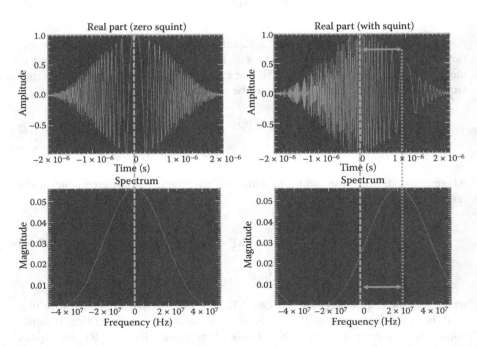

FIGURE 3.8 Squint effects on the Doppler shift.

For better inspection and to reveal the aliasing problem, we duplicate Figure 5.4 in [4] in Figure 3.9, showing the signal spectrum (real part) of a point target, a two-way antenna beam pattern as a function of θ_{diff}, and the aliases associated with the main compressed target. The decreasing power of the aliased targets (ghosts) is obvious for the weaker gain of the antenna side lobes. The spacing between the true target and those ghosts is about equal to the pulse period (pulse repetition interval [PRI]).

Figure 3.10 is an example of aliasing where target features along the coast repeatedly appear but with decreasing intensity along the azimuthal direction (indicated by arrows). This is exactly due to azimuth ambiguity.

When there are multiple targets randomly presented in the scene, the Doppler centroid is more uncertain in terms of its width and position. Figure 3.11 shows the resulting Doppler spreading effect, with one target at $t = 0$ s, five targets at $t = 1 \times 10^{-6}$ s, and two targets at $t = 1.5 \times 10^{-6}$ s, including a small squint effect. The presence of frequency spreading is caused by the mixing of the individual center frequencies corresponding to each target within a Doppler bandwidth. This poses a challenge to focusing a scene with a global Doppler centroid, as illustrated in Figure 3.12. Treatment of Doppler centroid estimation is discussed in Chapter 5.

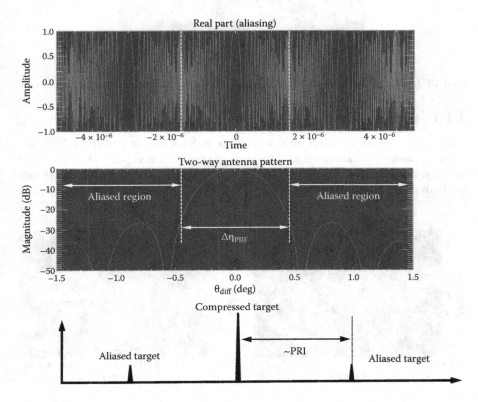

FIGURE 3.9 FM signal aliasing.

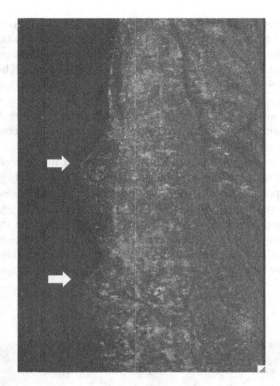

FIGURE 3.10   Azimuth ambiguity phenomena caused by signal aliasing (indicated by arrows).

3.3.3 PLATFORM VELOCITY

The platform velocity u plays an essential role in SAR system operation. From the radar equation, the slant range as a function of slow time is dependent on the radar velocity, Equation 3.4. As discussed earlier, the Doppler centroid and Doppler rate are both proportional to the squared velocity. Finally, the range cell migration that

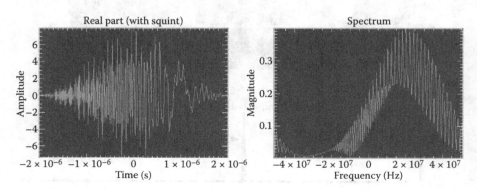

FIGURE 3.11   Effect of Doppler spreading.

Focused with global Doppler centroid   Focused with refined Doppler centroid

FIGURE 3.12 Effect of Doppler frequency and rate estimation on image focusing.

needs be corrected is inversely proportional to the squared velocity. Knowing the precise information of u is extremely critical in image focusing. However, it must differentiate the platform velocity and its ground velocity when the antenna beam sweeps across the target within its swath. As explained in [8], for the airborne system, because of the low flying altitude, the platform velocity is equal to the speed of the antenna beam sweeping across the target on the ground. In the spaceborne system, due to the earth's curvature effect and rotation and noncircular satellite orbit, the relative velocity is generally not constant and nonlinear. By taking these factors into account, the effective velocity is approximated as [4]

$$u \cong \sqrt{v_s v_g} \tag{3.20}$$

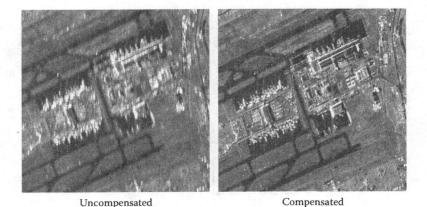

Uncompensated   Compensated

FIGURE 3.13 Platform velocity variation effect on image focusing.

where v_s is the platform velocity and v_g is the ground velocity; both vary with orbital vector, including the position and range (Figure 3.13). Consequently, the squint angle must be scaled from the physical squint angle according to [4]

$$\theta_r = \frac{u}{v_g}\theta_{sq} = \frac{v_s}{u}\theta_{sq} \tag{3.21}$$

3.4 FM CONTINUOUS WAVE

3.4.1 SIGNAL PARAMETERS

Referring to Figure 3.14, the transmitted signal is of the form [16,17]

$$s_t(t) = \exp\left\{ j2\pi\left(f_c t + \frac{1}{2}at^2 \right) \right\} \tag{3.22}$$

where a is the frequency change rate within the transmitted bandwidth B. With a, the intermediate frequency (IF) bandwidth B_{if} determines the range of the minimum and maximum delay times, τ_{min}, τ_{max}.

The maximum unambiguous range is determined by

$$R_{max} = \frac{f_{max}c}{2a} \tag{3.23}$$

where f_{max} is the corresponding maximum sampling frequency.

The received signal in one dimension is given by

$$s_r(\tau) = \exp\left\{ j2\pi\left[f_c(t-\tau) + \frac{1}{2}a(t-\tau)^2 \right] \right\} \tag{3.24}$$

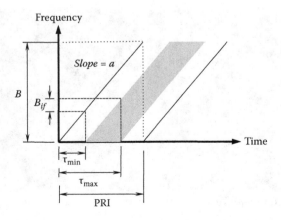

FIGURE 3.14 FMCW signal parameters.

The beat signal (intermediate frequency) for a single target takes the form

$$s_{if}(t) = s_t(t)s_r(t)^* = \exp\left\{ j2\pi\left[-f_c(t-\tau) - \frac{1}{2}a(t^2 - 2t\tau + \tau^2) + f_c t + \frac{1}{2}at^2 \right] \right\}$$

$$= \exp\left\{ j2\pi\left[f_c t + at\tau - \frac{1}{2}at^2 \right] \right\} \tag{3.25}$$

The delay time is range dependent; that is, it is subsequently a function of slow time:

$$\tau = \frac{2R(t,\eta)}{c} \tag{3.26}$$

$$R(t,\eta) = \sqrt{R_0^2 + u^2(t+\eta)^2} \tag{3.27}$$

For short range, the slant range may be approximated to

$$R(t,\eta) = \sqrt{R_0^2 + v^2(t+\eta)^2} \cong R_\eta + \frac{u^2\eta}{R_\eta}t \tag{3.28}$$

with

$$R_\eta = \sqrt{R_0^2 + (u\eta)^2} \tag{3.29}$$

Now Equation 3.25 may be written more explicitly, leading to

$$s_{if}(t,\eta) = \exp\left\{ j2\pi\left[f_c\left(\frac{2R_\eta}{c} - \frac{f_d\lambda}{c}t\right) + at\left(\frac{2R_\eta}{c} - \frac{f_d\lambda}{c}t\right) - \frac{2a}{c^2}R_\eta^2 \right.\right.$$

$$\left.\left. + \frac{2a}{c^2}R_\eta f_d\lambda t - \frac{af_d^2\lambda^2}{2c^2}t^2 \right] \right\} \tag{3.30}$$

where the range–Doppler frequency is induced by the time dependence in Equation 3.27 with

$$f_d = -\frac{2}{\lambda}\frac{\partial R(t,\eta)}{\partial t} = -\frac{2}{\lambda}\frac{u^2\eta}{R_\eta} \tag{3.31}$$

In focusing, it is essential to remove the Doppler frequency term by designing a proper filter [17–21]. The technical aspect is treated in Chapters 5 and 7.

3.4.2 OBJECT TO DATA MAPPING

The mapping from the object domain to the data domain is illustrated in Figure 3.15; t_0 is the near-range time. The range sampling frequency is confined in $f_r = \left[f_c - \dfrac{a}{2}, f_c + \dfrac{a}{2} \right)$, while the chirp rate is $a_r = \dfrac{4\pi}{c} \left[f_c - \dfrac{a}{2}, f_c + \dfrac{a}{2} \right)$. Note that $f_r = \alpha_{os,r}|a|$PRI. A simple simulation is carried out using the parameters given in Table 3.1. Two targets are located 750 m apart. The phase and amplitude of the IF signal is plotted in Figure 3.16, where the Fourier transform was performed on the range. Relevant focusing issues are discussed in Chapter 5.

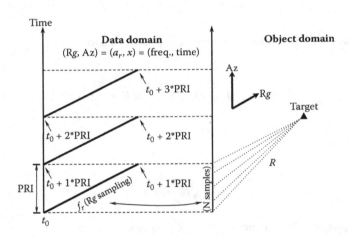

FIGURE 3.15   Mapping of the object domain and data domain.

TABLE 3.1
FMCW Signal Simulation Parameters for Point Targets

Item	Parameter	Value
Transmitted bandwidth	B	100 MHz
Carrier frequency	f_c	24 GHz
Pulse repeat interval	PRI	4.0 ms
Target location	R	250, 1000 m
Slant range	R_{range}	750~2000 m

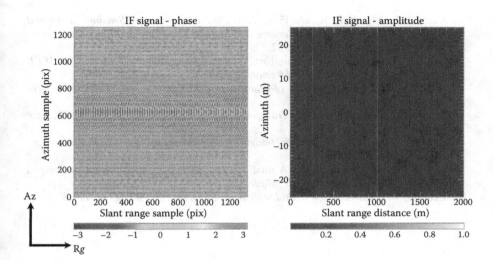

FIGURE 3.16 **(See color insert.)** Phase and amplitude of the IF signal after Fourier transform on the range using the parameters in Table 3.1.

REFERENCES

1. Barton, D. K., *Modern Radar System Analysis*, Artech House, Norwood, MA, 1988.
2. Berkowitz, R. S., ed., *Modern Radar: Analysis, Evaluation, and System Design*, John Wiley & Sons, New York, 1965.
3. Cook, C. E., *Radar Signals: An Introduction to Theory and Applications*, Artech House, Norwood, MA, 1993.
4. Cumming, I., and Wong, F., *Digital Signal Processing of Synthetic Aperture Radar Data: Algorithms and Implementation*, Artech House, Norwood, MA, 2004.
5. Curlander, J. C., and McDonough, R. N., *Synthetic Aperture Radar: Systems and Signal Processing*, Wiley-Interscience, New York, 1991.
6. Cantafio, L. J., *Space-Based Radar Handbook*, Artech House, Norwood, MA, 1989.
7. Pillai, S. U., Li, K. Y., and Himed, B., *Space Based Radar: Theory and Applications*, McGraw-Hill, New York, 2008.
8. Carrara, W. G., Majewski, R. M., and Goodman, R. S., *Spotlight Synthetic Aperture Radar: Signal Processing Algorithms*, Artech House, Norwood, MA, 1995.
9. Franceschetti, G., and Lanari, R., *Synthetic Aperture Radar Processing*, CRC Press, Boca Raton, FL, 1999.
10. Wahl, D. E., Eiche, P. H., Ghiglia, D. C., Thompson, P. A., and Jakowatz, C. V., *Spotlight-Mode Synthetic Aperture Radar: A Signal Processing Approach*, Springer, New York, 1996.
11. Brandwood, D., *Fourier Transforms in Radar and Signal Processing*, 2nd ed., Artech House, Norwood, MA, 2012.
12. Papoulis, A., *Fourier Integral and Its Applications*, McGraw-Hill, New York, 1962.
13. Soumekh, M., *Synthetic Aperture Radar Processing*, John Wiley & Sons, New York, 1999.
14. Stimson, G. W., *Introduction to Airborne Radar*, Scientific Publishing, Mendham, NJ, 1998.

15. Ulaby, F. T., Moore, R. K., and Fung, A. K., *Microwave Remote Sensing: Active and Passive*, vol. 2, *Radar Remote Sensing and Surface Scattering and Emission Theory*, Artech House, Norwood, MA, 1982.
16. Meta, A., Hoogeboom, P., and Ligthart, L., Signal processing for FMCW SAR, *IEEE Transactions on Geoscience and Remote Sensing*, 45(11): 3519–3532, 2007.
17. Richards, M. A., *Fundamentals of Radar Signal Processing*, 2nd ed., McGraw-Hill, New York, 2014.
18. Bracewell, R. M., *The Fourier Transform and Its Applications*, McGraw-Hill, New York, 1999.
19. Ferro-Famil, L., *Electromagnetic Imaging Using Synthetic Aperture Radar*, Wiley-ISTE, New York, 2014.
20. Gen, S. M., and Huang, F. K., A modified range-Doppler algorithm for de-chirped FM-CW SAR with a squint angle, *Modern Radar*, 29(11): 49–52, 2007.
21. Skolnik, M. I., *Introduction to Radar Systems*, McGraw-Hill, Singapore, 1981.

4 SAR Path Trajectory

4.1 INTRODUCTION

In synthetic aperture radar (SAR) data acquisition and processing, the Doppler frequency is one of the most vital parameters known for focusing the image. For SAR, the Doppler shift is determined by the radial velocity and is profoundly associated with the motion of the platform. This implies that the time and position of the platform must be known. Subsequently, it is important to understand to what extent the motion trajectory affects the image focusing. Recall that $f_d = -\frac{2}{\lambda}\frac{dR}{dt} = -\frac{2}{\lambda}u_r$. The radar's radial velocity u_r, the relative velocity between the sensor and the target, is determined from $u_r = -\vec{u} \cdot \hat{R}$, where $\vec{u}$ is the radar velocity and $\hat{R}$ is the unit range vector. For accurate azimuthal positioning of the target, it is critical and essential to estimate the position vectors of the radar and target as accurately as possible under a reference time and space coordinate systems. To uniquely describe and determine the SAR sensor position, velocity, and attitude requires a coordinate system or reference frame [1,2]. For a SAR system under study, we assume Newtonian mechanics is valid. Here the coordinate system refers to both time and space coordinates. For space coordinates, both satellite and airborne systems will be described in Chapters 5 and 6 to facilitate the discussions of SAR image focusing and motion compensation.

4.2 TIME COORDINATE SYSTEM

Before we introduce the orbital parameters, it is necessary to define the time system on which the coordinate can be defined. From [3], we have the following:

- *Julian Day (JD)*: Commonly used to calculate sidereal time when needed. Julian Day is a day as a unit, calculated from 12 midday January 1, 4713 BC to the present day. Hereafter the units (e.g., minutes and seconds) are based on a day unit conversion of a second kind. The Julian calendar was calculated correctly from ancient times to the present, on a 7-day week basis, and continued after a few days.
- *Universal Time (UT)*: Greenwich Meridian (0° longitude) at noon time.
- *Greenwich Mean Sidereal Time (GMST)*: Due to the angle variation of the earth's revolution around the sun every day and the precession and nutation effects, the actual sidereal time is not easy to calculate, so we use the mean sidereal time to replace it. If high-precision observation is not critical, it is more convenient to use mean sidereal time. It can be determined from the Julian Day by the following formula [3]:

$$\text{GMST} = 24110^{\text{s}}.54841 + 8640184^{\text{s}}.812866T_0 + 1.002737909350795\text{UT1}$$
$$+ 0^{\text{s}}.093104T^2 - 0^2.0000062T^3 \tag{4.1}$$

where

$$T_0 = \frac{\text{JD}(0^h\,\text{UT1}) - 2451545}{36525} \tag{4.2}$$

$$T = \frac{\text{JD(UT1)} - 2451545}{36525} \tag{4.3}$$

where the time unit is UT1 (another Universal Time [4]).

Greenwich Apparent Sidereal Time (GAST): Similar to GMS, but corrected by an average of the ecliptic (mean equinox) and drift from the equatorial plane (mean equator). So GAST is simply regarded as GMST plus a deviation related by

$$\text{GAST} = \text{GMST} + \Delta\psi\cos(\varepsilon) \tag{4.4}$$

where $\Delta\psi$, ε may be obtained from the International Earth Rotation and Reference Systems Service (IERS) Bulletin B form available at http://www.iers.org/.

Terrestrial Time (TT): Measured on the geoid with a constant offset to International Atomic Time (TAI):

$$\text{TT} = \text{TAI} + 32.154[\text{s}] \tag{4.5}$$

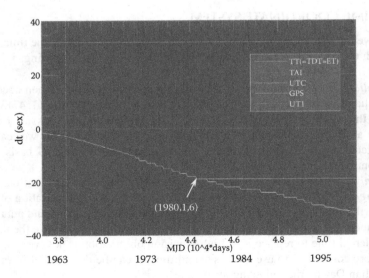

FIGURE 4.1 **(See color insert.)** Time differences of TT, TAI, UTC, GPS Time, and UT1 (from top to bottom).

Global Positioning System (GPS) Time: Differs from TAI by a constant amount:

$$GPS = TAI - 19[s] \qquad (4.6)$$

Each time definition as outlined above is not exactly the same; differences always exist among them. Following [4], Figure 4.1 plots the differences in seconds among the Terrestrial Time, International Atomic Time, Coordinated Universal Time (UTC), GPS Time, and Universal Time (UT1). For excellent treatment and details, refer to [4].

4.3 SPATIAL COORDINATE SYSTEM

4.3.1 Orbital Parameters

A satellite's orbital position as a function of time is most easily determined using orbital elements. Six fundamental parameters are commonly used to determine the orbital position: semimajor axis, eccentricity, inclination angle, right ascension of ascending node, argument of perigee, and rue anomaly (Figure 4.2). Table 4.1 gives the symbols and definitions of these parameters. Three of the parameters determine the orientation of the orbit or trajectory plane in space, while another three locate the body in the orbital plane [4–6]. These six parameters are uniquely related to the position and velocity of the satellite at a given epoch.

Kepler's equation relates the eccentric anomaly to the mean anomaly M as

$$M = E - e \sin E = \omega_{sn}(t - t_p) \qquad (4.7)$$

where E is the eccentric anomaly, t is time, t_p is the time of the satellite passage of periapse, and ω_{sn} is the mean motion, or the mean angular velocity of the satellite. The mean angular velocity of the satellite around one orbit is simply the orbital period divided by 2π:

$$\omega_{sn} = \frac{P}{2\pi} = \sqrt{\frac{a^3}{\mu}} \qquad (4.8)$$

Finally, the true anomaly can be computed from the eccentric anomaly as

$$\tan(\upsilon/2) = \sqrt{\frac{1+e}{1-e}} \tan(E/2) \qquad (4.9)$$

Note that of the six Keplerian elements, only the true anomaly v varies with time for two-body dynamics. The satellite range will be determined until the geocentric coordinate system is introduced.

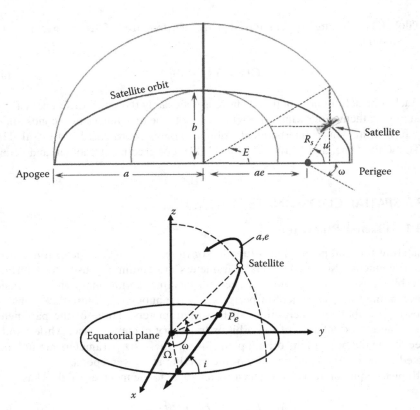

FIGURE 4.2 Keplerian orbital parameters.

TABLE 4.1
Definition of Orbital Parameters

Symbol	Definition
a	Semimajor axis
e	Eccentricity
i	Inclination angle
Ω	Right ascension of ascending node
ω	Argument of perigee
υ	True anomaly

4.3.2 GEOCENTRIC AND GEODETIC SYSTEMS AND THEIR TRANSFORMATION

Efforts have been devoted to establishing a global coordinate system. Various commonly used reference ellipsoids and data have been ever introduced. These systems are all originates at the earth's center [4,6–8]. The geodetic coordinate is supplementary to the Cartesian coordinate. In a geocentric system, the origin refers to the center of the earth, with the z-axis pointing to the North Pole. The geocentric system includes the earth central rotational (ECR) system and the earth central inertial (ECI) system. Shown in Figure 4.3, in ECR, the x-axis points to the Greenwich Meridian, the z-axis points to the North Pole, and the y-axis is perpendicular to the x- and z-axes. In ECI, the z-axis points to the North Pole, the x-axis points to the vernal equinox, and the y-axis is perpendicular to the x- and z-axes. In the figure, δ is the declination and α is the right ascension. Note that the xy-plane lies in the equatorial plane.

Referring to Figure 4.2, with semimajor axis a, semiminor axis b, and flatness $f = \dfrac{a-b}{a}$ as known parameters, the transformation from the geodetic latitude, the longitude, and the height above a reference ellipsoid φ, ϑ, h to the geocentric coordinate (x, y, z) takes the following forms [9–12]:

$$x = (N + h)\cos\varphi\cos\vartheta$$
$$y = (N + h)\cos\varphi\sin\vartheta \qquad (4.10)$$
$$z = [N(1 - e^2) + h]\sin\varphi$$

In the above equation,

$$N = \frac{a}{\sqrt{1 - e^2 \sin^2\varphi}} \qquad (4.11)$$

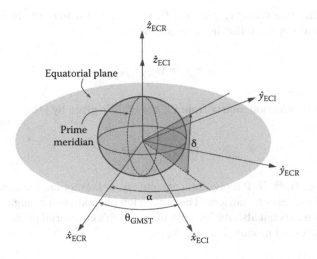

FIGURE 4.3 Earth central inertial and Earth central rotation coordinates.

where e is the eccentricity of the reference ellipsoid:

$$e = \sqrt{2f - f^2} \tag{4.12}$$

The inverse transformation of Equations 4.10 follows [13]

$$\varphi = 2 \tan^{-1}\left(\frac{z}{x + \sqrt{x^2 + y^2}}\right)$$

$$\vartheta = 2 \tan^{-1}\left(\frac{z}{D + \sqrt{D^2 + z^2}}\right) \tag{4.13}$$

$$h = \frac{\kappa + e^2 - 1}{\kappa}\sqrt{D^2 + z^2}$$

where

$$\kappa = \frac{h + N(1 - e^2)}{N} \tag{4.14}$$

$$D = \frac{\kappa\sqrt{x^2 + y^2}}{\kappa + e^2} \tag{4.15}$$

4.3.3 Conversion between ECR and ECI Coordinates

If we know the state vector $\mathbf{r}_{ECI}$ in the ECI system, its counterparts in the ECR system are obtained by the following transformation:

$$\mathbf{r}_{ECR} = \mathbf{U}_{ECR}^{ECI}\mathbf{r}_{ECI} \tag{4.16}$$

where the transformation matrix from ECI to ECR is given by [4]

$$\mathbf{U}_{ECR}^{ECI} = \Pi\Theta\mathbf{N}\mathbf{P} \tag{4.17}$$

where matrices Π, Θ, $\mathbf{N}$, $\mathbf{P}$ represent, respectively, polar motion, the earth's rotation, nutation, and precession matrices. The precession phenomenon changes not only the earth's rotation axis but also the location of the earth's equatorial plane. It is said that the y- and z-axes are modified according to

$$\mathbf{P}(T_1, T_2) = \mathbf{R}_z(-z(T, t))\mathbf{R}_y(\vartheta(T, t))\mathbf{R}_z(-\xi(T, t)) \tag{4.18}$$

where

$$\xi(T,t) = (2306''.2181 + 1''.39656T - 0''.00139T^2)t$$
$$+ (0''.30188 - 0''.000344T)t^2 + 0''017998t^3$$

$$z(T,t) = (2306''.2181 + 1''.39656T - 0''.00139T^2)t$$
$$+ (1''.09468 - 0''.000066T)t^2 + 0''018203t^3 \qquad (4.19)$$

$$\vartheta(T,t) = (2004''.3109 + 0''.85330T - 0''.00217T^2)t$$
$$+ (-0''.42665 - 0''.000217T)t^2 + 0''041833t^3$$

with t denoting the time difference $T_2 - T_1$,

$$t = T_2 - T_1 = \frac{JD_2(TT) - JD_1(TT)}{36525.0} \qquad (4.20)$$

with

$$T_1 = \frac{JD_1(UT1) - 2451545.0}{36525.0} \qquad (4.21)$$

Nutation occurs when no external torque (torque free) exists, but the body still rotates freely like a precession. However, the manner of the rotation is largely different from that of the precession by noting the fact that the angular momentum on each axis is not conserved, but the total angular momentum on the three axes is conserved. In celestial coordinates, nutation is generally ignored in defining the ecliptic and equatorial planes. Hence, strictly speaking, they should be called the mean ecliptic (mean equinox) and the mean equatorial (mean equator) plane. After nutation correction, they can be regarded as the correct coordinate plane while meeting the definition of instantaneous coordinates. The nutation matrix is

$$N(T) = R_x(-\varepsilon - \Delta\varepsilon)R_z(-\Delta\psi)R_z(\varepsilon) \qquad (4.22)$$

where ε is the angle between the mean equator and the mean equinox, ε' is the angle between the true equator and the true equinox, $\Delta\varepsilon = \varepsilon - \varepsilon'$ is the periodic change of the vernal equinox, and $\Delta\psi$ is the angle difference between the mean and the true equinox, or the periodic shift of the vernal equinox.

The sidereal time (or angle) is due to the angle change produced by the earth's revolution and rotation. This angle difference must be corrected when using the ECI or ECR coordinate system. This is done by the following earth rotation about the z-axis:

$$\Theta(t) = R_z(GAST). \qquad (4.23)$$

In Equation 4.23, is the Greenwich Apparent Sidereal Time (GAST), taking the form

$$GAST = GMST + \Delta\psi\cos(\varepsilon) \qquad (4.24)$$

with the Greenwich Mean Sidereal Time (GMST) given in Equation 4.1.

Aside from the precession and nutation, the earth also experiences polar motion due to the variations of the earth's mass distribution, which is season dependent. The accumulated polar motion components, x_p, y_p, are generally in meter scale; both are recorded in IERS Bulletin B available at http://www.iers.org/. The correction matrix is

$$\Pi(t) = \mathbf{R}_z(-x_p)\mathbf{R}_x(-y_p) \approx \begin{bmatrix} 1 & 0 & x_p \\ 0 & 1 & -y_p \\ -x_p & y_p & 1 \end{bmatrix} \qquad (4.25)$$

4.4 SATELLITE POSITION, VELOCITY, AND THEIR ESTIMATIONS

For a satellite platform, the orbital position as a function of time is the most easily determined using orbital elements.

Denoting the satellite position $\mathbf{R}_s$ and target position $\mathbf{R}_t$, both referring to the earth's center, the range between the satellite and target is (Figure 4.4)

$$R = |\mathbf{R}| = |\mathbf{R}_s - \mathbf{R}_t| \qquad (4.26)$$

If target velocity at $\mathbf{R}_t$ as measured on the earth surface is $\mathbf{u}_{rt}$, then $\dot{\mathbf{u}}_t = \vec{\omega}_e \times \mathbf{u}_t + \mathbf{u}_{rt}$.

The Doppler frequency induced by the moving of the platform, target, and earth may then be expressed as [9]

$$f_d = -\frac{2}{\lambda R}[\mathbf{u}_s \cdot (\mathbf{R}_s - \mathbf{R}_t) + \vec{\omega}_e \cdot (\mathbf{R}_s \times \mathbf{R}_t) - \mathbf{u}_{rt} \cdot (\mathbf{R}_s - \mathbf{R}_t)] \qquad (4.27)$$

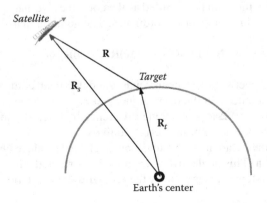

FIGURE 4.4 Simple geometry of a satellite SAR.

The Doppler rate is calculated by

$$
\begin{aligned}
d(R\dot{R})/dt = R\ddot{R} + \dot{R}^2 &= \mathbf{a}_s \cdot (\mathbf{R}_s - \mathbf{R}_t) + \mathbf{a}_s \cdot (\mathbf{u}_s - \mathbf{u}_t) \\
&+ \vec{\omega}_e \cdot [\mathbf{R}_s \times \mathbf{u}_t + (\mathbf{u}_s \times \mathbf{R}_t)] \\
&- \mathbf{u}_{rt} \cdot (\mathbf{u}_s - \mathbf{u}_t) - (\mathbf{R}_s - \mathbf{R}_t) \cdot (\mathbf{a}_{rt} + \vec{\omega}_e \times \mathbf{u}_{rt})
\end{aligned}
\tag{4.28}
$$

where $\mathbf{u}_t$, $\mathbf{a}_{rt}$ are the velocity and acceleration of the target relative to the earth's surface, respectively.

4.4.1 ORBITAL STATE VECTOR

It is essential to uniquely determine the satellite position, velocity, and epoch time in order to derive the Doppler frequency and Doppler rate as given by Equations 4.27 and 4.28. A numerical scheme can be devised to find the true anomaly υ, which is time dependent. This can be done by setting an initial guess of the eccentric anomaly E and iteratively refining the estimation of it until an error is within a preset bound. Equation 4.9 is used to relate the eccentric anomaly to the true anomaly. After numerically solving v, we can determine the satellite range R_s by the formula

$$
R_s = \frac{a(1 - e^2)}{1 + e \cos \upsilon}.
\tag{4.29}
$$

Then the position and velocity vectors in the perifocal frame are readily determined by (Figure 4.5)

$$
\mathbf{R}_{s(\mathrm{PER})} = \begin{bmatrix} R_s \cos \upsilon \\ R_s \sin \upsilon \\ 0 \end{bmatrix}
\tag{4.30}
$$

$$
\mathbf{u}_{(\mathrm{PER})} = \begin{bmatrix} -\sqrt{\mu/p} \sin \upsilon \\ \sqrt{\mu/p}(e + \cos \upsilon) \\ 0 \end{bmatrix}
\tag{4.31}
$$

Once the position and velocity vectors are defined in the perifocal frame, they may be rotated to the ECI frame through simple rotations of $-\Omega$ about the z-axis, $-\alpha_i$ about the x-axis, and $-\omega$ about the z-axis, as follows:

$$
\mathbf{R}_{s(\mathrm{ECI})} = \mathbf{R}_z(-\Omega)\mathbf{R}_x(-\alpha_i)\mathbf{R}_z(-\omega)\mathbf{R}_{s(\mathrm{PER})}
\tag{4.32}
$$

$$
\mathbf{u}_{(\mathrm{ECI})} = \mathbf{R}_z(-\Omega)\mathbf{R}_x(-\alpha_i)\mathbf{R}_z(-\omega)\mathbf{u}_{(\mathrm{PER})}
\tag{4.33}
$$

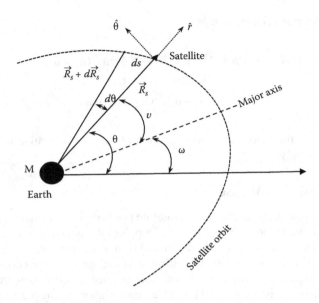

FIGURE 4.5 Satellite position and velocity.

where the rotation matrices about the x-, y-, and z-axes are

$$
\mathbf{R}_x(\Theta) = \begin{bmatrix} 1 & 0 & 0 \\ 0 & \cos\Theta & \sin\Theta \\ 0 & -\sin\Theta & \cos\Theta \end{bmatrix}
$$

$$
\mathbf{R}_y(\Theta) = \begin{bmatrix} \cos\Theta & 0 & -\sin\Theta \\ 0 & 1 & 0 \\ \sin\Theta & 0 & \cos\Theta \end{bmatrix} \tag{4.34}
$$

$$
\mathbf{R}_z(\Theta) = \begin{bmatrix} \cos\Theta & \sin\Theta & 0 \\ -\sin\Theta & \cos\Theta & 0 \\ 0 & 0 & 1 \end{bmatrix}
$$

To deduce the orbital elements of a satellite, at least six independent measurements are needed. In this chapter, we adopted simplified perturbation models (SPMs) [14–16], through two-line elements available at http://celestrak.com/, to estimate the orbital state vectors of the satellite of interest, for example, TerraSAR, in the ECI coordinate system. Once the orbit parameters are determined, we can then transform them to ECR coordinate system as described in Section 4.4.2.

The state vectors in the transformation are

$$
\mathbf{R}_{\mathrm{ECR}} = \mathbf{U}_{\mathrm{ECR}}^{\mathrm{ECI}}\, \mathbf{R}_{\mathrm{ECI}} \tag{4.35}
$$

$$\mathbf{u}_{\text{ECR}} = \mathbf{U}_{\text{ECR}}^{\text{ECI}} \mathbf{u}_{\text{ECI}} + \frac{d\mathbf{U}_{\text{ECR}}^{\text{ECI}}}{dt} \mathbf{R}_{\text{ECI}} \tag{4.36}$$

$$\mathbf{R}_{\text{ECI}} = \left(\mathbf{U}_{\text{ECR}}^{\text{ECI}}\right)^{\text{T}} \mathbf{R}_{\text{ECR}} \tag{4.37}$$

$$\mathbf{u}_{\text{ECI}} = \left(\mathbf{U}_{\text{ECR}}^{\text{ECI}}\right)^{\text{T}} \mathbf{u}_{\text{ECR}} + \frac{d\left(\mathbf{U}_{\text{ECR}}^{\text{ECI}}\right)^{\text{T}}}{dt} \mathbf{R}_{\text{ECR}} \tag{4.38}$$

where T denotes transpose operation. Note that from [9], in computing the time derivative of the transformations in Equation 4.36, the precession, nutation, and polar motion matrices may be considered constant for their small amount. Then, the time derivative of the transformation matrix is approximated as

$$\frac{d\mathbf{U}_{\text{ECR}}^{\text{ECI}}}{dt} \approx \Pi \frac{d\Theta}{dt} \mathbf{NP} \tag{4.39}$$

4.4.2 RADAR BEAM-POINTING VECTOR

The satellite velocity vector can be put into the form

$$\mathbf{u}_0 = [0, 0, u_s]^{\text{T}} \tag{4.40}$$

where u_s is the tangent velocity. The radar beam-pointing vector for a look angle θ_ℓ to the target is expressed as

$$\mathbf{p}_0 = [-\cos(\theta_\ell), \sin(\theta_\ell), 0]^{\text{T}} \tag{4.41}$$

In the above equation, the radar beam-pointing vector is defined in its own coordinate. Now it is necessary to transform to the ECR coordinate. This requires the satellite's yaw and pitch angles, $\theta_{\text{yaw}}, \theta_{\text{pitch}}$. The transformation matrix is

$$\mathbf{M}_{01} = \mathbf{R}_{\mathbf{x}}(\theta_{\text{yaw}})\mathbf{R}_{\mathbf{y}}(\theta_{\text{pitch}}) \tag{4.42}$$

where the rotation matrix is defined in Equation 4.34. Now, expressed in the ECR coordinate, we have

$$\mathbf{u}_1 = \mathbf{u}_0$$
$$\mathbf{p}_1 = \mathbf{M}_{01}\mathbf{p}_0 \tag{4.43}$$

The transformation from ECR to the earth-centered orbit plane (ECOP) coordinate transform is through

$$\mathbf{u}_2 = \mathbf{M}_{12}\mathbf{u}_1$$
$$\mathbf{p}_2 = \mathbf{M}_{12}\mathbf{p}_1$$

(4.44)

where the transformation matrix is

$$\mathbf{M}_{12} = \mathbf{R}_y(\theta_{\text{hour}})$$

(4.45)

with θ_{hour} denoting the hour angle measured from the ascending node.

If θ_{incl}, the orbit plane inclination angle, is known, we can transform the satellite velocity and beam-pointing vectors to the ECI coordinate by the following relations:

$$\mathbf{u}_3 = \mathbf{M}_{23}\mathbf{u}_2$$
$$\mathbf{p}_3 = \mathbf{M}_{23}\mathbf{p}_2$$

(4.46)

The transformation matrix $\mathbf{M}_{23}$ is constructed as

$$\mathbf{M}_{23} = \mathbf{R}_x(\theta_{\text{incl}})$$

(4.47)

Sometimes it is desired to perform local latitude correction between the geodetic and geocentric coordinates. This is easily done by means of the following conversion:

$$\mathbf{u}_{3g} = \mathbf{M}_{3g}\mathbf{u}_3$$
$$\mathbf{p}_{3g} = \mathbf{M}_{3g}\mathbf{p}_3$$

(4.48)

The transformation matrix is $\mathbf{M}_{3g}$

$$\mathbf{M}_{3g} = \mathbf{R}_z^T\left(\theta_{lon}\right)\mathbf{R}_y(\delta\varphi_{lat})\mathbf{R}_z(\theta_{lon})$$

(4.49)

where ϑ_{lon} is the longitude; $\delta\varphi_{lat} = \varphi_{gd} - \varphi_{gc}$ is the difference between the geodetic latitude and the geocentric latitude with φ_{gd} denoting geodetic latitude and φ_{gc} the geocentric latitude.

4.4.3 Target Position on Earth

For a given or known satellite position and the radar beam-pointing vector determined by the above steps, it is possible to find the target position on the same coordinate system so that the satellite, radar, and target are all collocated in time and space.

$$\mathbf{R}_t = \mathbf{M}_{st}\left(\mathbf{R}_s, \mathbf{p}_{3g}\right)$$

(4.50)

where $\mathbf{M}_{st}$ is the transformation matrix from satellite positions on the earth's surface that follow the beam direction. This requires the target's center longitude and latitude. The transformation takes the forms

$$\mathbf{TTM}_{north} = \mathbf{R}_z(\alpha_t)\mathbf{R}_y(\theta_Y)\mathbf{R}_z(180°)\mathbf{R}_z(\vartheta_t)$$

$$\mathbf{TTM}_{south} = \mathbf{R}_z(\alpha_t)\mathbf{R}_y(\theta_Y)\mathbf{R}_z(\vartheta_t)$$

$$(4.51)$$

where α_t is the target's aspect angle measured counterclockwise from true north, $\mathbf{TTM}_{north}$ is for the target in the northern hemisphere, $\mathbf{TTM}_{south}$ is for the target in the southern hemisphere, and

$$\theta_Y := -\text{sgn}(\varphi_t) \times (90° + |\varphi_t|)$$

where sgn(·) is the sign function and φ_t, ϑ_t are the target's latitude and longitude.

The final stage is to line up the target's aspect angle to the radar beam-pointing direction.

We summarize the coordinate transformations of the satellite position vector and velocity, radar beam-point vector, and target position vector in the following steps:

1. Use SGP/SDP4 from two-line elements to find the satellite state vector for a time duration.
2. Input the polar motion, earth rotation, nutation, and precession matrices from IERS Bulletin B, followed by transforming the estimated state vector from ECI to ECR coordinates.
3. Given the satellite altitude, transform the satellite velocity and radar beam-pointing vectors from the local satellite coordinate to the earth's center.
4. Transform the above vectors to the ECOP.
5. Transform the resulting vectors to ECI.
6. Correct the local latitude between the geodetic and geocentric systems, followed by conversing the beam-pointing vector from ECI to ECR.
7. With the satellite state vector and beam-pointing vector, locate the target center position vector on the earth's surface in the ECR coordinate.
8. Transform from ECR to the geodetic system.
9. Transform from the geodetic to the geocentric system to find the target center's latitude, longitude, and height (φ, ϑ, h).
10. Given the aspect angle and the above target center, find the position vector in the ECR coordinate for each scatter (with possible different RCS values) within the target.

Figure 4.6 schematically details the above steps in converting the satellite state vector, radar beam-pointing vector, and target position vector coordination.

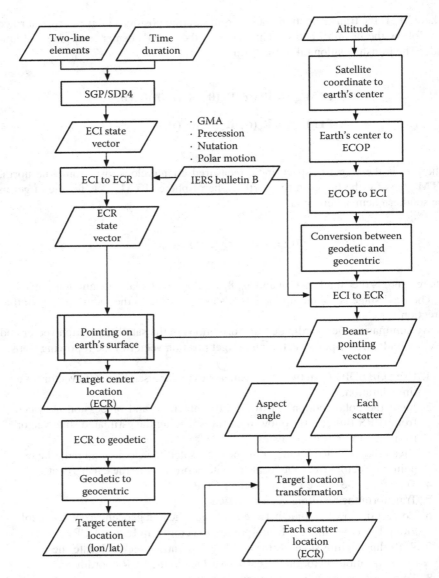

FIGURE 4.6 Flowchart of the coordinate conversions for the satellite state vector, radar beam-pointing vector, and target locations.

4.5 COORDINATES FOR AIRCRAFT PLATFORM

Now that the satellite coordinate system has been introduced, this section describes the common way to determine the platform position and altitude with respect to the ECR system for SAR operation that is mounted on a fixed-wing aircraft. Figure 4.7 depicts the roll, pitch, yaw coordinate system [5,16–22]. The origin is referred to as the aircraft center of gravity.

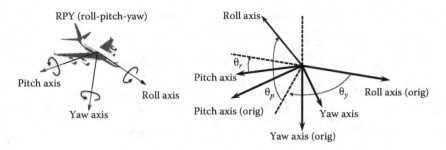

FIGURE 4.7 Definition of the RPY coordinate.

The rotation sequence may use the axis-angle method or the Euler angle method. In axis-angle method or the intrinsic rotation, we have

1. **Yaw** (RH for down, LH for up)
2. **Pitch** (RH)
3. **Roll** (RH)

In the Euler angle method or the extrinsic rotation, we have

1. **Yaw** (RH for down, LH for up)
2. **Pitch** (RH)
3. **Roll** (RH)

Note that the angle from counterclockwise (CCW) is designated as positive.

- **Intrinsic** rotation: Rotating with local axis (upper)
- **Extrinsic** rotation: Rotating with global axis (bottom)

4.5.1 ENU and NED Coordinates

Two types of local tangent planes (LTPs) are commonly used to specify the aircraft's altitude and velocity: ENU (east, north, up) and NED (north, east, down) (Figure 4.8) [22]. The conversion matrices from ENU to NED and vice versa are given by

$$\mathbf{C}_{NED}^{ENU} = \mathbf{C}_{ENU}^{NED} = \begin{vmatrix} 0 & 1 & 0 \\ 1 & 0 & 0 \\ 0 & 0 & -1 \end{vmatrix} \quad (4.52)$$

Given a user's local coordinate, x_u, y_u, z_u, the transformation from the ECR to the ENU coordinate is

$$\mathbf{X}_{ENU} = \mathbf{C}_{ENU}^{ECR} \mathbf{X}_{ECR} + \tilde{\mathbf{r}} \quad (4.53)$$

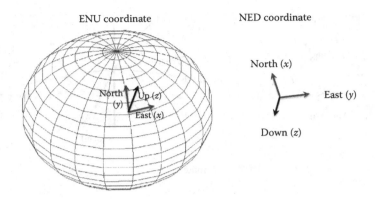

FIGURE 4.8 ENU and NED coordinates.

where $\tilde{\mathbf{r}}$ is the coordinate origin shift vector from ECR to the local reference, and φ_ℓ, ϑ_ℓ are the local latitude and longitude, respectively.

$$\tilde{\mathbf{r}} = \begin{bmatrix} x_u \sin \vartheta_\ell - y_u \cos \vartheta_\ell \\ x_u \sin \varphi_\ell \cos \vartheta_\ell - y_u \sin \varphi_\ell \cos \vartheta_\ell - z_u \cos \varphi_\ell \\ -x_u \cos \varphi_\ell \cos \vartheta_\ell - y_u \cos \varphi_\ell \sin \vartheta_\ell - z_u \sin \varphi_\ell \end{bmatrix} \qquad (4.54)$$

The transformation matrix in Equation 4.53 is given by

$$\mathbf{C}_{\text{ENU}}^{\text{ECR}} = \begin{bmatrix} -\sin \vartheta_\ell & \cos \vartheta_\ell & 0 \\ -\sin \varphi_\ell \cos \vartheta_\ell & -\sin \varphi_\ell \sin \vartheta_\ell & \cos \varphi_\ell \\ \cos \varphi_\ell \cos \vartheta_\ell & \cos \varphi_\ell \sin \vartheta_\ell & \sin \varphi_\ell \end{bmatrix} \qquad (4.55)$$

The transformation from ENU back to the ECR coordinate in Equation 4.53 is formulated as

$$\mathbf{X}_{\text{ECR}} = \left[\mathbf{C}_{\text{ENU}}^{\text{ECR}} \right]^{-1} (\mathbf{X}_{\text{ENU}} - \tilde{\mathbf{r}}) \qquad (4.56)$$

4.5.2 RPY COORDINATE AND FLIGHT PATH ANGLE

Consider the radar beam pointing to the target. It is necessary to integrate the sensor position and altitude first and then to transform to the ECR coordinate, where the

FIGURE 4.9 Flight path angle γ.

position involves the ENU/NED, while altitude is transformed from the roll, pitch, yaw (RPY) to the ENU/NED coordinate by the following transformation matrix:

$$\mathbf{C}_{ENU}^{RPY} = \begin{bmatrix} \sin(\theta_y)\cos(\theta_p) & \cos(\theta_p)\cos(\theta_y)+\sin(\theta_r)\sin(\theta_y)\sin(\theta_p) & -\sin(\theta_r)\cos(\theta_y)+\cos(\theta_r)\sin(\theta_y)\sin(\theta_p) \\ \cos(\theta_y)\cos(\theta_p) & -\cos(\theta_r)\sin(\theta_y)+\sin(\theta_r)\cos(\theta_y)\sin(\theta_p) & \sin(\theta_r)\sin(\theta_y)+\cos(\theta_r)\cos(\theta_y)\sin(\theta_p) \\ \sin(\theta_p) & -\sin(\theta_r)\cos(\theta_p) & -\cos(\theta_r)\cos(\theta_p) \end{bmatrix}$$

$$(4.57)$$

As shown in Figure 4.9, the flight path angle γ is the angle between the velocity vector (flight path) and the local horizon, which describes whether the aircraft is climbing or descending. Note that the flight path angle is independent of the aircraft altitude.

4.5.3 SIMULATION EXAMPLE

In the airborne SAR system, platform motion poses a challenge as far as image focusing is concerned. This section provides a simulation example of how the aircraft motion is evaluated.

In the coordinate system in Figure 4.10, $\left(\hat{d}_l, \hat{d}_\perp, \hat{d}_\parallel\right)$ represents the original line-of-sight, perpendicular, velocity (LPV) coordinate, $\left(\hat{d}_{los}^{new}, \hat{d}_\perp^{new}, \hat{d}_{vel}^{new}\right)$ the original LPV

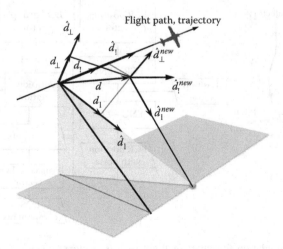

FIGURE 4.10 Definition of LPV coordinate.

coordinate with position bias, and $(d_{los}, d_\perp, d_{vel})$ the position bias component with d as the absolute total position bias. Note that

$$\hat{d}_\perp = \hat{d}_{los} \times \hat{d}_{vel}, \quad \hat{d}_{vel} \neq \hat{d}_\perp \times \hat{d}_{los}$$
$$\hat{d}_\perp^{new} = \hat{d}_{los}^{new} \times \hat{d}_{vel}^{new}, \quad \hat{d}_{vel}^{new} \neq \hat{d}_\perp^{new} \times \hat{d}_{los}^{new} \tag{4.58}$$

In what follows, we illustrate the path variations due to position and altitude noise under a SAR geometry. Figure 4.11 gives the complete flowchart for the simulation, with numerical parameters given in Table 4.2. The input and output parameters are listed and explained below. In Chapter 6, the path simulation will be used to study the motion effects and their compensation in regard to image focusing. Hence, the simulation provides a useful tool in designing and evaluating the focusing and motion compensation algorithms.

Input parameters:
- Initial position: The start and final point longitude, latitude, and height with geocentric coordinate
- Reference ellipsoid: WGS84
- SAR geometry: Look and squint angles
- Position noise: ENU coordinate position noise
- Altitude (roll, pitch, and yaw) noise: Altitude noise (roll → pitch → yaw axis)

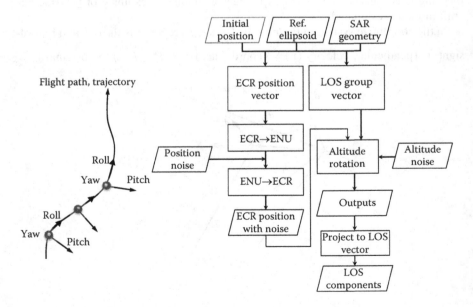

FIGURE 4.11 Simulation flowchart of flight path (trajectory) given a noisy position and altitude of the aircraft. The reference ellipsoid is WGS84.

TABLE 4.2

Simulation Parameters of Flight Path Trajectory

Item	Values
Initial Position	
Start position (Lon, Lat, H) (deg, deg, km)	(120.3891807, 24.1213241, 10.0)
End position (Lon, Lat, H) (deg, deg, km)	(120.3786004, 24.2442277, 10.5)
SAR Geometry	
Look angle [deg]	30
Squint angle [deg]	0
Reference Ellipsoid	
Semi-major axis [m]	6378137 (WGS84)
Semi-minor axis [m]	6356752.31414 (WGS84)
Number of simulation sample	10
Noise Error or Bias	
Simulation position noise (ENU) (mean, std) (m, m)	(0, 100)
Simulation attitude angle noise (RPY) (mean, std) (deg, deg)	(0, 2.5)

- Noise: White Gaussian noise (WGN) with specified mean and standard deviation:
  - ENU position noise (WGN, zero mean, standard = constant)
  - Altitude angle noise (WGN, zero mean, standard = constant)

Output parameters:
  - ENU position
  - ENU position (noise)
  - ECR position (altitude)
  - ECR position (altitude and noise)
  - New altitude axis unit vector (RPY axis)
  - Original line of sight (LOS) group vector (LOS vector, LOS perpendicular vector, tangent velocity vector)
  - LOS group vector with altitude rotated (altitude)

Distance project on LOS components:
  - Vector component of error or bias projected on new LOS group vector

Figure 4.12 displays three ENU components for noise-free, noise, and resulting noise. Similar simulations are carried out for the RPY components. With the SAR geometry parameters in Table 4.1, the resulting variation in slant range is illustrated in Figure 4.13, with an ideal slant range, noisy components, and the difference between the ideal and noisy slant ranges. It is this differential component that causes the problem in Doppler estimation and is the term that must be precisely estimated in order to correct the Doppler information. Figure 4.14 displays the variations of slant range due to noisy flight path from starting and ending positions (Table 4.1). Random and instable fluctuations of slant ranges are very visible compared to the ideal and

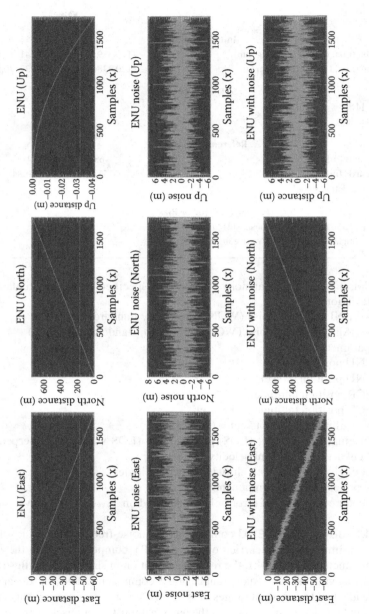

FIGURE 4.12 Ideal ENU components (top), noisy components (middle), and resulting noisy ENU components (bottom).

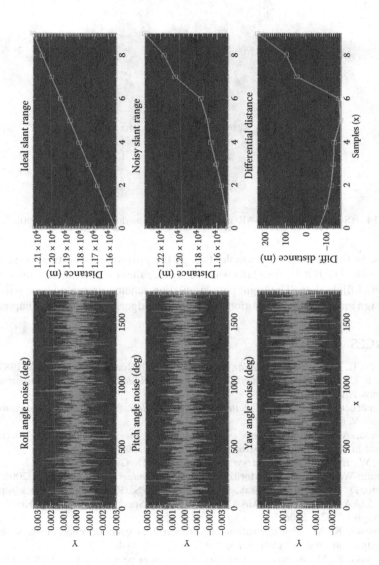

FIGURE 4.13 RPY noise (left column), ideal slant range (top, right), noisy range (middle), and differential slant range caused by the RPY noise.

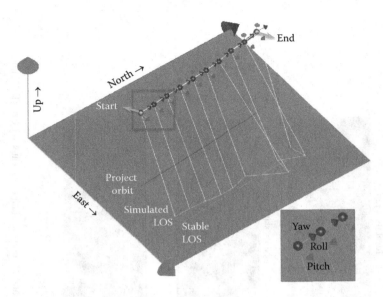

FIGURE 4.14 **(See color insert.)** SAR slant range variations due to noisy flight path.

stable ones. In this simulation example, though the position and altitude noise are generated with WGN, it is straightforward to replace them with real measured and recorded POS flight data. The flight path simulation scheme presented here will be used to design and evaluate the motion compensation algorithms given in Chapter 6.

REFERENCES

1. Cantafio, L. J., *Space-Based Radar Handbook*, Artech House, Norwood, MA, 1989.
2. Pillai, S. U., Li, Y. K., and Himed, B., *Space Based Radar: Theory and Applications*, McGraw-Hill, New York, 2008.
3. Montenbruck, O., and Gill, E., *Satellite Orbits: Models, Methods, and Applications*, Springer Verlag, Berlin, 2000.
4. Chatfield, A. B., *Fundamentals of High Accuracy Inertial Navigation*, American Institute of Aeronautics and Astronautics, Reston, VA, 1997.
5. Torge, W., and Muller, J., *Geodesy: An Introduction*, De Gruyter, Berlin, 2012.
6. Hofmann-Wellenhof, B., and Moritz, H., *Physical Geodesy*, Springer, Berlin, 2006.
7. Vallado, D., Crawford, P., Sujsak, R., and Kelso, T. S., Revisiting spacetrack report no. 3, AIAA 2006-6753, AIAA/AAS Astrodynamics Specialist Conference, Keystone, CO, August 21–24, 2006.
8. Borkowski, K. M., Transformation of geocentric to geodetic coordinates without approximations, *Astrophysics and Space Science*, 139: 1–4, 1987.
9. Borkowski, K. M., Accurate algorithms to transform geocentric to geodetic coordinates, *Bulletin Geodesy*, 63: 50–56, 1989.
10. Bowring, B. R., Transformation from spatial to geodetic coordinates, *Survey Review*, 23: 323–327, 1976.
11. Bowring, B. R., The accuracy of geodetic latitude and height equations, *Survey Review*, 38: 202–206, 1985.

12. Vermeille, H., Direct transformation from geocentric coordinates to geodetic coordinates, *Journal of Geodesy*, 76: 451–454, 2000.
13. Curlander, J. C., and McDonough, R. N., *Synthetic Aperture Radar: Systems and Signal Processing*, Wiley-Interscience, New York, 1991.
14. Miura, N. Z., Comparison and design of simplified general perturbation models (SGP4) and code for NASA Johnson Space Center, Orbital Debris Program Office, MS thesis, California Polytechnic State University, San Luis Obispo, 2009.
15. Noerdlinger, P., Theoretical basis for the SDP toolkit geolocation package for the ECS project, Technical report, Hughes Applied Information System, Landover, MD, 1995.
16. Farrell, J. A., *The Global Positioning System and Inertial Navigation*, McGraw-Hill, New York, 1998.
17. Farrell, J. A., *Aided Navigation: GPS with High Rate Sensors*, McGraw-Hill, New York, 2008.
18. Grewal, M. S., Weill, L. R., and Andrews, A. P., *Global Positioning Systems, Inertial Navigation, and Integration*, 2nd ed., Wiley, New York, 2007.
19. Hofmann-Wellenhof, B., Lichtenegger, H., and Wasle, E., *GNSS—Global Navigation Satellite Systems: GPS, GLONASS, Galileo, and More*, Springer, New York, 2007.
20. Noureldin, A., Karamat, T. B., and Georgy, J., *Fundamentals of Inertial Navigation, Satellite-Based Positioning and Their Integration*, Springer, Berlin, 2013.
21. Rogers, R. M., *Applied Mathematics in Integrated Navigation Systems*, American Institute of Aeronautics and Astronautics, Reston, VA, 2003.
22. Stevens, B. L., and Lewis, F. L., *Aircraft Control and Simulation*, John Wiley & Sons, New York, 1992.
23. Hedgley, D. R., Jr., *An Exact Transformation from Geocentric to Geodetic Coordinates for Nonzero Altitudes*, NASA TR R-458, National Aeronautics and Space Administration, Washington, DC 1976.
24. Hoots, F. R., and Roehrich, R. L., Models for propagation of NORAD element sets, Spacetrack report no. 3, Department of Defense, Washington, DC, 1980.

5 SAR Image Focusing

5.1 INTRODUCTION

This chapter mainly deals with image formation or focusing. We present in a systems point of view, along with numerical simulation, the exploration of synthetic aperture radar (SAR) imaging properties that are profoundly associated with platform motion and trajectory, as presented in the previous chapter. Two major issues closely embracing image focusing are particularly addressed: estimation of Doppler centroid and estimation of Doppler rate [1–9]. Motion compensation that, in most cases, is necessary to refine the focusing quality will be treated in Chapter 6 in more detail. Commonly used focusing algorithms, the range–Doppler algorithm (RDA) and chirp scaling algorithm (CSA) are illustrated through numerical simulations. Of these algorithms, each bears its own advantages and shortcomings, with some suitable for large swaths and others better able to handle greater fluctuating sensor trajectories, such as an airborne system. Since the mentioned algorithms are well documented in the literature [10–17], it would be exhaustive and redundant to repeat descriptions of the algorithms here. Instead, for the purpose of this book, we will only focus on the RDA and CSA algorithms. For other algorithms, such as the $\omega - k$ method and focusing on the spotlight imaging mode, useful references are given in [10,11]. A modified RDA for the frequency-modulated continuous-wave (FMCW) system was presented in [18]. The technical issues addressed in RDA and CSA are equally and equivalently applied to the rest of the available methods.

5.2 GENERIC RANGE–DOPPLER ALGORITHM

The range–Doppler algorithm (RDA) is conceptually straightforward for converting the object to the image in the range–Doppler plane, at which the SAR operates.

Recall from Equation 3.1 that the received echo signal at the baseband takes the following form:

$$s_0(\tau, \eta) = A_0 p_r\left(\tau - \frac{2R(\eta)}{c}\right) g_a(\eta - \eta_c) \exp\left\{-j4\pi f_c \frac{R(\eta)}{c}\right\}$$
$$\times \exp\left\{j\pi a_r\left(\tau - \frac{2R(\eta)}{c}\right)^2\right\}$$

(5.1)

Putting together, the phase term in Equation 5.1 is given by

$$\theta(\tau) = -4\pi f_c \frac{R(\eta)}{c} + \pi a_r\left(\tau - \frac{2R(\eta)}{c}\right)^2 - 2\pi f_\tau \tau$$

(5.2)

Upon applying the principle of the stationary phase, the fast time and range frequency can be approximately related by

$$\frac{d\theta}{d\tau} = 2\pi a_r \left(\tau - \frac{2R(\eta)}{c} \right) - 2\pi f_\tau = 0 \rightarrow \tau = \frac{f_\tau}{a_r} + \frac{2R(\eta)}{c} \tag{5.3}$$

We give the principle of the stationary phase in the appendix for ease of reference.

Substituting the fast time in Equation 5.3 into Equation 5.2, and taking the Fourier transform in the range direction, we obtain [19]

$$S_0(f_\tau, \eta) = \int_{-\infty}^{\infty} s_0(\tau, \eta) \exp\{-j2\pi f_\tau \tau\} d\tau = A_0 A_1 \tilde{p}_r(f_\tau) g_a(\eta - \eta_c)$$
$$\times \exp\left\{-j \frac{4\pi(f_c + f_\tau)}{c} R(\eta)\right\} \exp\left\{-j\pi \frac{f_\tau^2}{a_r}\right\} \tag{5.4}$$

where the last term on the right-hand side is informationless, undesired, and therefore is to be removed from range compression. Notice that the range frequency must obey the sampling theory, namely, $f_\tau \in \left[\dfrac{-f_s}{2}, \dfrac{f_s}{2}\right]$ with the analog-to-digital conversion (ADC) sampling frequency. In practice, the range frequency is much lower than the carrier frequency, that is, $f_c + f_\tau \approx f_c$.

Now, we take the Fourier transform of Equation 5.4 in the azimuth:

$$S_0(f_\tau, f_\eta) = \int_{-\infty}^{\infty} S_0(f_\tau, \eta) \exp\left\{-j2\pi f_\eta \eta\right\} d\eta$$

$$= A_0 A_1 A_2 \tilde{p}_r(f_\tau) G_a(f_\eta - f_{\eta_c}) \exp\{j\theta_a(f_\tau, f_\eta)\} \tag{5.5}$$

where the azimuth antenna pattern, expressed in the azimuth frequency domain, is

$$G_a(f_\eta) = g_a \left(\frac{-cR_0(f_\eta)}{2(f_c + f_\tau)u^2 \sqrt{1 - \dfrac{c^2(f_\eta)^2}{4u^2(f_c + f_\tau)^2}}} \right) \tag{5.6}$$

The phase term, also a slow time dependence, is

$$\theta_a(\eta) = -\frac{4\pi(f_c + f_\tau)R(\eta)}{c} - \frac{\pi f_\tau^2}{a_r} - 2\pi f_\eta \eta \tag{5.7}$$

Similar to the treatment of the phase term in Equation 5.2, applying the principle of the stationary phase, we can decouple η and f_η:

$$\frac{\partial \theta_a(\eta)}{\partial \eta} = 0 \Rightarrow f_\eta = -\frac{2u^2(f_c + f_\tau)\eta}{c\sqrt{R_0^2 + u^2\eta^2}} \Rightarrow \eta = -\frac{cR_0 f_\eta}{2(f_c + f_\eta)u^2\sqrt{1 - \frac{c^2 f_\eta^2}{4u^2(f_c + f_\tau)^2}}} \qquad (5.8)$$

With the approximate relation of η and f_η, the approximate phase becomes

$$\hat{\theta}_a(f_\tau, f_\eta) = -\frac{4\pi R_0(f_c + f_\tau)}{c}\sqrt{1 - \frac{c^2 f_\eta^2}{4V_r^2(f_c + f_\tau)^2}} - \frac{\pi f_\tau^2}{a_r}$$

$$= -\frac{4\pi R_0 f_c}{c}\sqrt{D^2(f_\eta, u) + \frac{2f_\tau}{f_c} + \frac{f_\tau^2}{f_c^2}} - \frac{\pi f_\tau^2}{a_r} \qquad (5.9)$$

It is convenient to define [12]

$$D(f_\eta, u) = \sqrt{1 - \frac{c^2 f_\eta^2}{4u^2 f_c^2}} = \sqrt{1 - \frac{\lambda^2 f_\eta^2}{4u^2}} \qquad (5.10)$$

$$Z(R_0, f_\eta) = \frac{cR_0 f_\eta^2}{2u^2 f_c^3 D^3(f_\eta, u)} \qquad (5.11)$$

Now, substituting the phase of Equation 5.9 into Equation 5.5, and taking the inverse Fourier transform on the range, the result is

$$S_{rd}(\tau, f_\eta) = \int_{-\infty}^{\infty} S_0(f_\tau, f_\eta)\exp\{j2\pi f_\tau \tau\} df_\tau$$

$$= A_0 A_1 A_2 A_3 p_r \left[a_m \left(\tau - \frac{2R_0}{cD(f_\eta, u)} \right) \right] G_a \left(f_\eta - f_{\eta_c} \right) \qquad (5.12)$$

$$\times \exp\left\{ j\frac{4\pi R_0 f_c D(f_\eta, u)}{c} \right\} \exp\left\{ j\pi a_m \left[\tau - \frac{2R_0}{cD(f_\eta, u)} \right]^2 \right\}$$

where we can see that the chirp rate modified by the factor becomes

$$a_m = \frac{a_r}{1 - Z a_r} = \frac{a_r}{1 - a_r \left(\dfrac{cR_0 f_\eta^2}{2u^2 f_c^3 D^3(f_\eta, u)} \right)} \qquad (5.13)$$

and the range dependence, with modification factor $D(f_\eta, u)$, becomes

$$R_{rd}(R_0, u) = \frac{R_0}{D(f_\eta, u)} \tag{5.14}$$

The slow time dependence range, accounted by $D(f_\eta, u)$, causes range cell migration (RCM)—the echo energy spreads along the azimuth direction (see Figure 3.6).

The approximate of phase term $\hat{\theta}_a(f_\tau, f_\eta)$ in Equation 5.9 must now be made when transforming back to the fast time domain $\hat{\theta}_a(\tau, f_\eta)$. To do so, we expand it into Taylor series:

$$\hat{\theta}_a(f_\tau, f_\eta) = -\frac{4\pi R_0 f_c}{c} \sqrt{D^2(f_\eta, u) + \frac{2f_\tau}{f_c} + \frac{f_\tau^2}{f_c^2} - \frac{\pi f_\tau^2}{a_r}}$$

$$\approx -\frac{4\pi R_0 f_c}{c} \left[D^2(f_\eta, u) + \frac{f_\tau}{f_c D(f_\eta, u)} - \frac{c^2 f_\eta^2}{4u^2 f_c^2} \frac{f_\tau^2}{2 f_c^2 D^3(f_\eta, u)} \right] - \frac{\pi f_\tau^2}{a_r} \tag{5.15}$$

Again, applying the principle of the stationary phase to obtain the relation of the azimuth frequency,

$$\frac{\partial \theta(f_\tau)}{\partial f_\tau} = -\frac{4\pi R_0}{c D(f_\eta, u)} + 2\pi Z f_\tau - \frac{2\pi f_\tau}{a_r} + 2\pi\tau = 0 \Rightarrow f_\eta = a_m \left[\tau - \frac{2R_0}{c D(f_\eta, u)} \right] \tag{5.16}$$

With the azimuth frequency f_η given above, the phase in Equation 5.9 can now be written in more explicit form:

$$\hat{\theta}_a(\tau, f_\tau) = \hat{\theta}_a(f_\tau, f_\eta) + 2\pi f_\tau \tau$$

$$= -\frac{4\pi R_0 f_c}{c} \left[D^2(f_\eta, u) + \frac{f_\tau}{f_c D(f_\eta, u)} - \frac{c^2 f_\eta^2}{4u^2 f_c^2} \frac{f_\tau^2}{2 f_0^2 D^3(f_\eta, u)} \right] \tag{5.17}$$

$$- \frac{\pi f_\tau^2}{a_r} + 2\pi f_\tau \tau$$

From Equations 5.16 and 5.17, Equation 5.12 represents the results of the range–Doppler signal and may be put into a more compact form as

$$S_{rd}(\tau, f_\eta) = A' G_a(f_\eta - f_{\eta_c}) A_{rcm} \exp\{\phi_\tau\} \exp\{\phi_{f_\eta}\} \tag{5.18}$$

where the phase term, a slow time dependent, is recognized as

$$\phi_\tau = \exp\left\{ j\pi a_m \left[\tau - \frac{2R_{rd}}{c} \right] \right\} \tag{5.19}$$

which is to be corrected in range compression; the phase term below is to be removed by azimuth compression:

$$\phi_{f_\eta} = \exp\left\{ j \frac{4\pi R_0 f_c D(f_\eta, u)}{c} \right\} \tag{5.20}$$

The spreading of signal energy across the range while the sensor is moving is

$$A_{rcm} = p_r \left[a_m \left(\tau - \frac{2R_{rd}}{c} \right) \right] \tag{5.21}$$

Physically, $Z(R_0, f_\eta)$ in Equation 5.11 serves as an additional chirp rate in the second range compression, designated as $a_{src}(R_0, f_\eta)$. So, the modified chirp rate can be expressed as

$$a_m = \frac{a_r}{1 - a_r / a_{src}} \tag{5.22}$$

Now that the range compression is performed by multiplying Equation 5.18, a counterphase term of Equation 5.19 is

$$S_{rc}(\tau, f_\eta) = A_0 p_r \left(\tau - \frac{2R_{rd}(f_\eta)}{c} \right) G_a(f_\eta - f_{n_c}) \exp\left\{ -j \frac{4\pi f_c R_0}{c} \right\} \exp\left\{ \frac{f_\eta}{a_a} \right\} \tag{5.23}$$

where the Doppler rate a_a can be found in Equation 5.8. Inclusion of $a_{src}(R_0, f_\eta)$ in range compression is called second range compression in [20–22]. If motion error is nonnegligible or even profound, a second range compression is desirable to refine the focusing performance at the stage of range compression.

5.2.1 RANGE CELL MIGRATION CORRECTION

The time-dependent range R_{rd} in Equation 5.14 can be explicitly written as

$$R_{rd} = \frac{R_0}{D(f_\eta, u)} = R_0 + R_0 \left\{ \frac{1}{D(f_\eta, u)} - 1 \right\} = R_0 + R_0 \left\{ \frac{1 - D(f_\eta, u)}{D(f_\eta, u)} \right\} \triangleq R_0 + \Delta R_{rcm} \tag{5.24}$$

By expansion of $D(f_\eta, u)$ above, the range cell migration ΔR_{rcm} takes the expression

$$\Delta R_{rcm} = R_0 \left\{ \frac{\lambda^2 f_\eta^2}{8u^2} + \frac{3}{8} \left(\frac{\lambda^2 f_\eta^2}{u^2} \right)^2 + \frac{5}{16} \left(\frac{\lambda^2 f_\eta^2}{u^2} \right)^3 + \cdots \right\} \tag{5.25}$$

If the squint angle is small, the first term of Equation 5.24 is dominant and thus must be included. In a high-squint or strong motion system, higher-order terms beyond the quadratic term are desirable or even necessary.

After correcting the range cell migration, if all goes well, the signal in Equation 5.23 becomes

$$S_{rmc}(\tau, f_\eta) = A_0 p_r \left(\tau - \frac{2R_0}{c} \right) G_a \left(f_\eta - f_{\eta_c} \right) \exp \left\{ -j \frac{4\pi f_0 R_0}{c} \right\} \exp \left\{ j\pi \frac{f_\eta^2}{a_a} \right\}$$

(5.26)

The final step is to perform inverse fast Fourier transform (IFFT) in the azimuth to get back to the range–azimuth domain:

$$s_{rc}(\tau, \eta) = \int_{-\infty}^{\infty} \left\{ S_{rmc}(\tau, f_\eta) \exp \left[j \frac{4\pi R_0}{\lambda} + j\pi \frac{f_\eta^2}{a_a} \right] \right\} \exp\{ j2\pi f_\eta \eta \} \, df_\eta$$

(5.27)

At this point, three approximations are made if the squint angle is small. The range cell migration is simplified as

$$\Delta R_{rcm} = R_0 \left[\frac{1 - D(f_\eta, u)}{D(f_\eta, u)} \right] \Rightarrow \Delta R_{rck} \approx \frac{\lambda^2 R_0 f_\eta^2}{8u^2}$$

(5.28)

The modified chirp rate is simply replaced by the chirp rate; if R_0, u, f_η are all constant, a range compression filter is simplified by

$$H_{rg}(f_\tau) = \exp \left\{ j\pi \frac{f_\tau^2}{a_m(R_0, f_\eta)} \right\} \Rightarrow H_{rg}(f_\tau) = \exp \left\{ j\pi \frac{f_\tau^2}{a_r} \right\}$$

(5.29)

If R_0 is constant at the midrange, then we can multiply the range frequency signal by a secondary range compression phase:

$$H_{src}(f_\tau) = \exp \left\{ -j\pi \frac{f_\tau^2}{a_{src}(R_0, f_\eta)} \right\}$$

(5.30)

Table 5.1 summarizes the chirp rate, range compression filter, range cell migration corrector, and azimuth compression filter for the cases of low-squint and high-squint angles. The use of approximate filters and a corrector involves less computational cost, but at the price of degraded focusing performance. Figure 5.1 illustrates the effect of the range cell migration correction (RCMC) using Radarsat-1 data provided in [12]. Generally, RCMC is required even for a stable orbital SAR system like Radarsat-1.

TABLE 5.1
Comparison of Approximations Made for Low-Squint and High-Squint Cases

Item	Low Squint	High Squint
Chirp rate	a_r	a_m
Range compression filter	$\exp\left\{ j\pi \dfrac{f_\tau^2}{a_r} \right\}$	$\exp\left\{ j\pi \dfrac{f_\tau^2}{a_r} \right\} \exp\left\{ j\pi \dfrac{f_\tau^2}{a_{scr}(R_0, f_n)} \right\}$
Range cell migration correction	$R_{rcmc}(f_\eta) = \dfrac{\lambda^2 R_0 f_\eta^2}{8u^2}$	$R_{rcmc}(f_\eta, u) = R_0 \left[\dfrac{1 - D(f_\eta, u)}{D(f_\eta, u)} \right]$
Azimuth compression filter	$\exp\left\{ -j\pi \dfrac{f_\eta^2}{a_a} \right\}$	$\exp\left\{ j\pi \dfrac{4\pi R_0 D(f_\eta, u) f_c}{c} \right\}$

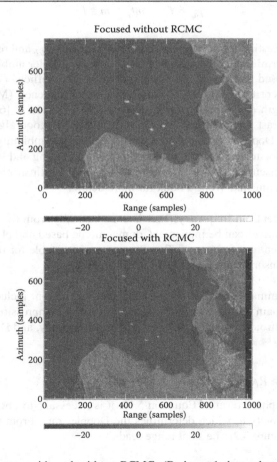

FIGURE 5.1 Images with and without RCMC. (Radarsat-1 data taken from Cumming, I. G., and Wong, F. H., *Digital Processing of Synthetic Aperture Radar Data: Algorithms and Implementation*, Artech House, Norwood, MA, 2005.)

5.2.2 Doppler Centroid Estimation

It is essential to estimate the Doppler centroid within the resolution cell, which occupies a certain Doppler bandwidth. Recall that

$$f_{\eta_c} = -\frac{2u_r}{\lambda} = -\frac{2u^2\eta_c}{\lambda R(\eta_c)} = \frac{2u\sin\theta_{r,c}}{\lambda} \qquad (5.31)$$

Therefore, finding the Doppler centroid requires an estimate of the sensor radial velocity of the scene.

The Doppler frequency is confined by the pulse repetition frequency (PRF), or $f_{\eta_c} \in [-f_p/2, f_p/2]$. However, due to the wraparound of the PRF, Equation 5.31 ends up as

$$f_{\eta_c} = f_{dcb} + mf_p, \quad m \in I \qquad (5.32)$$

This involves the estimates of baseband centroid frequency f_{dcb} and resolves the PRF ambiguity—determining correct m. To do so, various Doppler ambiguity resolvers have been proposed, such as angle of cross-correlation coefficient (ACCC) [9,14], multilook cross-correlation (MLCC), multilook beat frequency (MLBF) [3,4,12], and the wavelength diversity algorithm (WDA) [2]. As stated in [6], WDA works well in low-contrast scenes, while MLCC and MLBF are good algorithm choices that estimate the Doppler centroid and resolve the Doppler ambiguity. Various techniques are well documented in an excellent book by Cumming and Wong [12]. Here we only briefly discuss them for the purpose of numerical illustration. Readers are referred to Cumming and Wong's book for more details.

- Baseband centroid frequency: This is done basically from the echo signal. Two alternatives can be implemented: magnitude based and phase based.
- Absolute centroid frequency: Two sources are attainable for this information: the sensor's state vector and the echo signal.

Figure 5.2 summarizes a flowchart of Doppler estimation, including frequency and rate, in the chain of image focusing. Before the range compression, estimation of the baseband centroid frequency, absolute centroid frequency, and Doppler ambiguity must be done.

5.2.3 Doppler Rate Estimation

In removing the phase term of Equation 5.12, it is necessary to obtain an estimate of the sensor velocity u, which determines the Doppler rate. From the observation geometry (see Figure 3.2), the slant range reads

$$R(\eta) = \sqrt{R_0^2 + u^2\left(\eta - \eta_0'\right)^2} \qquad (5.33)$$

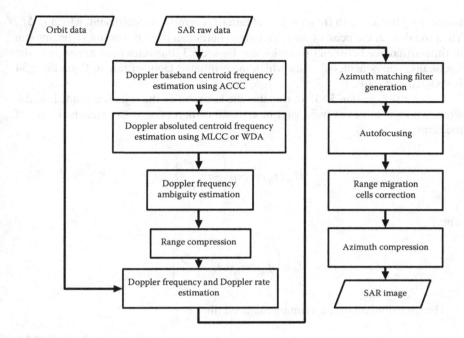

FIGURE 5.2 Flowchart of Doppler frequency and Doppler rate estimation.

Notice that the shortest range is a function of look angle θ_ℓ.

Once the velocity is known, the Doppler rate is computed as

$$\hat{a}_a = \frac{2u^2 \cos^3(\theta_q)}{\lambda R_0} \tag{5.34}$$

We may further refine the estimation by means of autofocusing, as shown in the estimation flowchart of Figure 5.2.

Let the Doppler rate be represented as a sum of a mean component and a fluctuation component:

$$\hat{a}_a = \bar{a}_a + \delta a_a \tag{5.35}$$

and

$$\delta a_a \approx -\hat{a}_a^2 \frac{\Delta \eta}{\Delta f_\eta} \tag{5.36}$$

where $\Delta \eta$ is the time difference between two looks and Δf_η is the separated frequency between two looks. In practice, we take the cross-correlation between two looks along the azimuth frequency to obtain the correlation coefficient, which is Δf_η. Then the shift of the peak of the correlation coefficient from the origin is the amount of time difference between two looks, $\Delta \eta$. Figure 5.3 illustrates the estimation of the fluctuation component, δa_a, with which the estimated Doppler rate in Equation 5.34 is updated and refined.

Inspecting Equation 5.13, we find that the last term on the right-hand side is undesired, which is to be removed later in azimuth compression with a matched filter of the form

$$H_{az}(f_\eta) = \exp\left\{ -j\pi \frac{f_\eta^2}{a_a} \right\} \tag{5.37}$$

where

$$f_\eta \in \left[f_{\eta_c} - \frac{f_p}{2}, f_{\eta_c} + \frac{f_p}{2} \right]$$

The time domain of the azimuth matched filter is

$$h_{az}(\eta) = \exp\{-j\pi a_a \eta^2\} \tag{5.38}$$

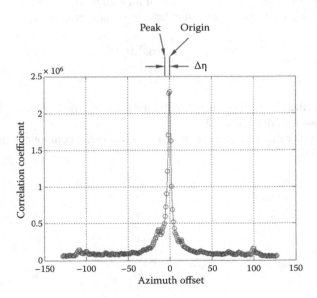

FIGURE 5.3 Estimation of $\Delta \eta$ from cross-correlation of two looks along the azimuth frequency.

where

$$\eta \in \left[-\frac{T_a}{2}, \frac{T_a}{2} \right]$$

See Figure 1.6.

The results after azimuth compression are

$$s_{ac}(\tau, \eta) = \int_{-\infty}^{\infty} \left\{ S_{rmc}(f_\tau, \eta) \cdot \exp\left[-j\pi \frac{f_\eta^2}{a_a} \right] \right\} \exp\{ j2\pi f_\eta \eta \} \, df_\eta$$

$$= A_0 p_r \left(\tau - \frac{2R_0}{c} \right) g_a(\eta) \exp\left\{ -j \frac{4\pi f_c R_0}{c} \right\} \exp\{ j2\pi f_{\eta_c} \eta \}$$

(5.39)

In the above expression, we readily recognize that the term $\exp\left\{ -j \frac{4\pi f_c R_0}{c} \right\}$ contains the target position information, while the phase term, $\exp\{ j2\pi f_{\eta_c} \eta \}$, is induced by the nonzero Doppler centroid f_{η_c}. For the purpose of illustration, we present numerical simulations for the case of a point target using the sensor parameters listed in Table 5.2.

TABLE 5.2
Sensor Parameters for Simulation of Range–Doppler Focusing

Parameter	Symbol	Numeric Value	Unit
Slant range of scene center (Target location)	$R(n_c)$	20	km
Effective radar velocity	u_r	150	m/s
Transmitted pulse duration	T_r	2.5	μs
Range FM rate	a_r	20×10^{12}	Hz/s
Radar center frequency	f_c	5.3	GHz
Doppler bandwidth	B_d	80	Hz
Range sampling rate	f_s	60	MHz
PRF	f_p	100	Hz
Number of range lines (Line)	N_a	256	pixels
Samples per range line (Sample)	N_r	128	pixels
Beam squint angle	$\vartheta_{r,c}$	0(Test)/3.5 (low)/21.9 (High)	deg
Beam center crossing time	n_c	0/−8.1/−49.7	s
Doppler centroid frequency	$f_{n,c}$	0/320/1975	Hz

In the following, we present the intermediate results of the range compression in the slant range–azimuth domain and slant range–Doppler domain, as shown in Figure 5.4. The range migration cells and their correction are shown in Figure 5.5.

Comparing the results of Figures 5.4 and 5.5, the effect of range cell migration and the corrected results is clearly seen. Notice that the frequency response is confined to the PRF range, as shown in Figure 5.6. The range cell migration

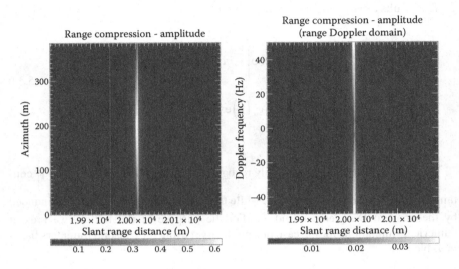

FIGURE 5.4 Results of the range compression in the slant range–azimuth domain (left) and slant range–Doppler domain (right).

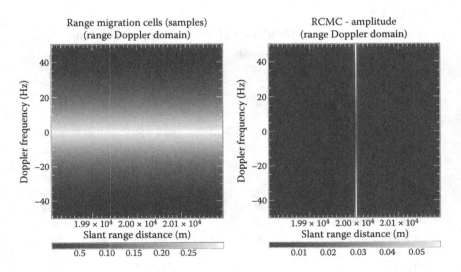

FIGURE 5.5 Range migration cells (left) and the results of their correction in the range–Doppler domain (right).

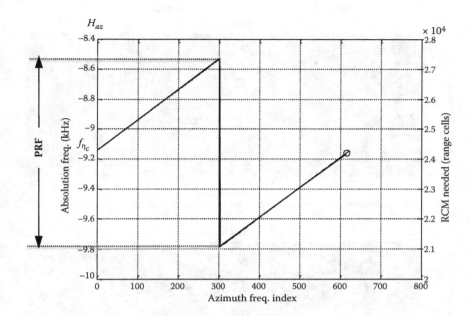

FIGURE 5.6 Azimuth matched filter for azimuth compression.

in correspondence to the matched filter is given in Equation 5.11. Figure 5.7 is a focused image of a point target after completing the range–Doppler algorithm as described above.

While the RDA procedure as outlined above seems quite straightforward, care must be taken in designing the azimuth matched filter given in Equation 5.37

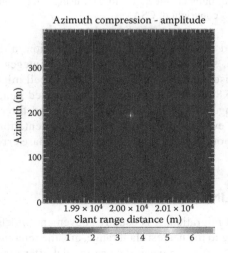

FIGURE 5.7 **(See color insert.)** Focused image (amplitude) after azimuth compression.

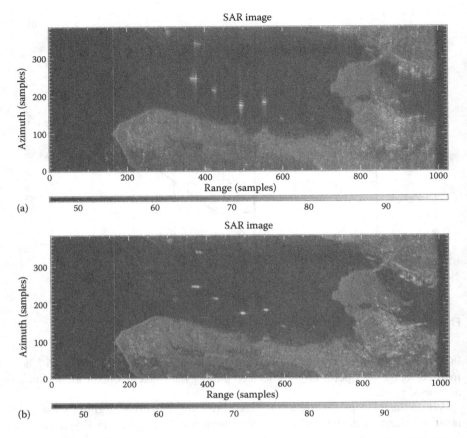

FIGURE 5.8 Sampling number effects on RCMC using integer number (a) and floating number (b). SAR images copyright of CSA.

or 5.38. It is realized that in numerical implementation, a discrete form of the filter is used. Hence, sampling number must be noninteger, for example, floating number. Otherwise, accurate correction of range cell migration is not always guaranteed. Figure 5.8 displays a comparison of focused images using an integer and floating sampling number in the azimuth matched filter to compensate the phase deviation. It is evident that in numerical implementation, a floating number is always required in order to accurately remove the phase error induced by range migration.

5.3   CHIRP SCALING ALGORITHM

In correcting the cell migration, it is usually convenient to define the azimuth frequency f_{η_c} as referring to the nearest slant range, and the reference range as referring to the scene center R_{ref}. A more general signal model after range compression may be expressed as

$$S_{rd}(\tau, f_\eta) = Ap_r \left[a_m \left(\tau - \frac{2R_0}{cD(f_\eta, u)} \right) \right] G_a(f_\eta - f_{\eta_c})$$

$$\times \exp\left\{ j \frac{4\pi R_0 D(f_\eta, u) f_c}{c} \right\} \exp\left\{ j\pi a_m \left[\tau - \frac{2R_0}{cD(f_\eta, u)} \right]^2 \right\} \tag{5.40}$$

5.3.1 RANGE CELL MIGRATION

For the moment, with the aid of $D(f_\eta, u)$ in Equation 5.10, we may decompose the total range cell migration into bulk and differential components [12,20–22], as given in the following equations. The reason for this shall become clear in later discussion.

$$\text{RCM}_{\text{total}}(R_0, f_\eta) \equiv R_{rd}(R_0, f_\eta) - R_{rd}(R_0, f_{\eta_{\text{ref}}}) \approx \frac{R_0}{D(f_\eta, u)} - \frac{R_0}{D\left(f_{\eta_{\text{ref}}}, u\right)} \tag{5.41}$$

$$\text{RCM}_{\text{bulk}}(f_\eta) \equiv \text{RCM}_{\text{total}}(R_{\text{ref}}, f_\eta) \approx \frac{R_{\text{ref}}}{D(f_\eta, u)} - \frac{R_{\text{ref}}}{D\left(f_{\eta_{\text{ref}}}, u\right)} \tag{5.42}$$

$$\text{RCM}_{\text{diff}}(R_0, f_\eta) \equiv \text{RCM}_{\text{total}}(R_0, f_\eta) - \text{RCM}_{\text{bulk}}(f_\eta)$$

$$\approx \left[\frac{R_0}{D(f_\eta, u)} - \frac{R_0}{D\left(f_{\eta_{\text{ref}}}, u\right)} \right] - \left[\frac{R_{\text{ref}}}{D(f_\eta, u)} - \frac{R_{\text{ref}}}{D\left(f_{\eta_{\text{ref}}}, u\right)} \right] \tag{5.43}$$

Figure 5.9 shows the relation of these three components.

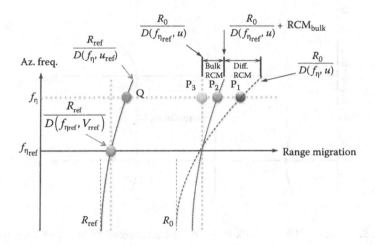

FIGURE 5.9 Relation of total RCM, bulk RCM, and differential RCM components in constructing the scaling function.

The bulk RMC is a series of duplications along the range direction. If we subtract the bulk RMC from the total RMC, the remaining is the residual or differential RMC, which is usually ignored in the generic RDA algorithm. Further, to remove it is a finer correction. The fast time of the range at point P_2 in Figure 5.9 is

$$\tau = \frac{2}{c} \left\{ \frac{R_0}{D(f_{\eta_{ref}}, u)} + \left[\frac{R_{ref}}{D(f_\eta, u_{r_{ref}})} - \frac{R_{ref}}{D(_{\eta_{ref}}, u_{r_{ref}})} \right] \right\} \qquad (5.44)$$

Now we replace the fast time by the reference range time

$$\tau' = \tau - \frac{2R_{ref}}{cD(f_\eta, u_{r_{ref}})} \qquad (5.45)$$

The time difference due to the differential RCM is computed as

$$\Delta\tau = \frac{2}{c} \mathrm{RCM}_{diff}(R_0, f_\eta) \qquad (5.46)$$

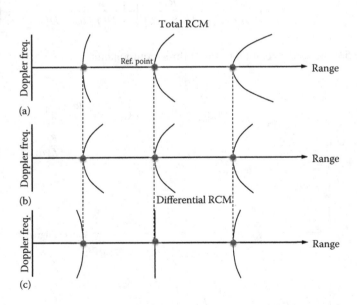

FIGURE 5.10 Total RCM (a), bulk RCM (b), and differential RCM (c), showing the order of correction.

or

$$\Delta\tau(\tau', f_\eta) = \left[\frac{D(f_{\eta_{ref}}, u)}{D(f_\eta, u)} - 1 \right] \tau' + \frac{2R_{ref}}{c} \left[\frac{D(f_{\eta_{ref}}, u)}{D(f_{\eta_{ref}}, u_{ref})D(f_\eta, u)} - \frac{1}{D(f_\eta, u_{ref})} \right]$$

(5.47)

Before proceeding, we illustrate the total RCM, bulk RCM, and differential RCM in the range–azimuth frequency domain (Figure 5.10), showing their order of corrections to be made.

5.3.2 CHIRP SCALING FUNCTIONS

A chirp scaling, a quadratic phase function, can be defined [12,20,21] to account for the time difference in Equation 5.47:

$$y_{sc}(R_0, f_\eta) = RCM_{diff}(R_0, f_\eta) \frac{2a_m}{c} = a_m \Delta\tau$$

(5.48)

and the phase of the scaling function should be of the form

$$\psi_{sc}(\tau', f_\eta) = \exp\left\{ j2\pi \int_0^{\tau'} a_m \Delta\tau(\xi, f_\eta) d\xi \right\}$$

(5.49)

Assuming that the sensor velocity, u, and the modified chirp rate, a_m, are range invariant, the second term in the differential fast time $\Delta\tau$ (Equation 5.47) may be ignored. The phase of the scaling function is approximated to

$$\psi_{sc}(\tau', f_\eta) = \exp\left\{ j\pi a_m \left[\frac{D(f_{\eta_{ref}}, u_{\tau_{ref}})}{D(f_\eta, u_{ref})} - 1 \right] (\tau')^2 \right\}$$

(5.50)

Accordingly, the scaling function is constructed as

$$S_{sc}(\tau', f_\eta) = \exp\{ j\psi_{sc}(\tau', f_\eta) \} = \exp\left\{ j\pi a_m \left[\frac{D(f_{\eta_{ref}}, u)}{D(f_\eta, u)} - 1 \right] (\tau')^2 \right\}$$

(5.51)

We multiply the range–Doppler signal by the scaling function

$$S_1(\tau, f_\eta) = S_{rd}(\tau, f_\eta) S_{sc}(\tau', f_\eta) \tag{5.52}$$

Taking the Fourier transform of the range direction (fast time), we have

$$S_2(f_\tau, f_\eta) = \int_{-\infty}^{\infty} S_1(\tau, f_\eta) \exp\{-j2\pi f_\tau \tau\} d\tau$$

$$= A_1 \tilde{p}_r(f_\tau) G_a \left(f_\eta - f_{\eta_c} \right) \exp \left\{ -j \frac{4\pi R_0}{cD\left(f_{\eta_{ref}}, u_{r_{ref}} \right)} f_\tau \right\}$$

$$\exp \left\{ -j \frac{4\pi R_0 f_c D(f_\eta, u)}{c} \right\}$$

$$\exp \left\{ -j \frac{4\pi a_m}{c^2} \left[1 - \frac{D\left(f_\eta, u_{r_{ref}} \right)}{D\left(f_{\eta_{ref}}, u_{r_{ref}} \right)} \right] \left[\frac{R_0}{D(f_\eta, u)} - \frac{R_{ref}}{D(f_\eta, u)} \right]^2 \right\} \tag{5.53}$$

$$\exp \left\{ -j \frac{\pi D(f_\eta, u)}{D\left(f_{\eta_{ref}}, u \right)} f_\tau^2 \right\} \exp \left\{ -j \frac{4\pi}{c} \left[\frac{1}{D\left(f_\eta, u_{r_{ref}} \right)} - \frac{1}{D\left(f_{\eta_{ref}}, u_{r_{ref}} \right)} \right] R_{ref} f_\tau \right\}$$

Now it is recognized that the first phase term is the desired term that contains the target location, the second and third exponential terms are the secondary range compression (SRC) and RCM, which will be removed in the second scaling process later, and the fourth, the azimuth modulation, and fifth exponential terms are to be removed in azimuth compression. Putting Equation 5.53 into more compact form,

$$S_{2df}(f_\tau, f_\eta) = \int_{-\infty}^{\infty} S_0(f_\tau, \eta) \exp\{-j2\pi f_\eta \eta\} d\eta$$

$$= A_0 A_1 A_2 \tilde{p}_r(f_\tau) G_a \left(f_\eta - f_{\eta_c} \right) \exp\{j\theta_a(f_\tau, f_\eta)\} \tag{5.54}$$

Notice that the antenna pattern in the azimuth frequency domain is

$$G_a(f_\eta) = g_a \left(\frac{-cR_0(f_\eta)}{2(f_c + f_\tau)u^2 \sqrt{1 - \dfrac{c^2(f_\eta)^2}{4u^2(f_c + f_\tau)^2}}} \right) \tag{5.55}$$

The coupling of f_c and f_η in the antenna pattern showing the above also explains the occurrence of range migration.

The phase term in Equation 5.54 is

$$\theta_a(f_\tau, f_\eta) = -\frac{4\pi R_0(f_c + f_\tau)}{c}\sqrt{1 - \frac{c^2 f_\eta^2}{4u^2(f_c + f_\tau)^2}} - \frac{\pi f_\tau^2}{a_r}$$

$$= -\frac{4\pi R_0 f_c}{c}\sqrt{D^2(f_\eta, u) + \frac{2f_\tau}{f_c} + \frac{f_\tau^2}{f_c^2}} - \frac{\pi f_\tau^2}{a_r}$$

(5.56)

Notice that in deriving the explicit phase term, we make use of the principle of the stationary phase in relating η, f_η, arriving at the following relation:

$$\eta = -\frac{cR_0 f_\eta}{2(f_c + f_\eta)u^2 \sqrt{1 - \frac{c^2 f_\eta^2}{4u^2(f_c + f_\tau)^2}}}$$

(5.57)

Now the second scaling function is constructed as

$$S_3(f_\tau, f_\eta) = S_2(f_\tau, f_\eta) S_{2df}(f_\tau, f_\eta)$$

(5.58)

or more explicitly,

$$S_3(f_\tau, f_\eta) = A_1 \tilde{p}_r(f_\tau) G_a\left(f_\eta - f_{\eta_c}\right) \exp\left\{-j\frac{4\pi R_0 f_c D(f_\eta, u)}{c}\right\}$$

$$\times \exp\left\{-j\frac{4\pi a_m}{c^2}\left[1 - \frac{D\left(f_\eta, u_{ref}\right)}{D\left(f_{\eta_{ref}}, u_{ref}\right)}\right]\left[\frac{R_0}{D(f_\eta, u)} - \frac{R_{ref}}{D(f_\eta, u)}\right]^2\right\}$$

(5.59)

Taking the inverse Fourier transform of Equation 5.59, we obtain the expression in range–azimuth frequency domain:

$$S_4(\tau, f_\eta) = \int_{-\infty}^{\infty} S_3(f_\tau, f_\eta) \exp\{j2\pi f_\tau \tau\} d\tau$$

$$= A_2 p_r\left(\tau - \frac{2R_0}{cD\left(f_{\eta_{ref}}, u_{ref}\right)}\right) G_a\left(f_\eta - f_{\eta_c}\right)$$

(5.60)

$$\times \exp\left\{-j\frac{4\pi R_0 f_c D(f_\eta, u)}{c}\right\}$$

$$\times \exp\left\{-j\frac{4\pi a_m}{c^2}\left[1 - \frac{D\left(f_\eta, u_{ref}\right)}{D\left(f_{\eta_{ref}}, u_{ref}\right)}\right]\left[\frac{R_0}{D(f_\eta, u)} - \frac{R_{ref}}{D(f_\eta, u)}\right]^2\right\}$$

The azimuth filter must have frequency response

$$\hat{H}_{az}(R_0, f_\eta) = \exp\left\{ j \frac{4\pi R_0 f_c D(f_\eta, u_r)}{c} \right\} \tag{5.61}$$

The residual phase error corrector is

$$\hat{H}_{rp}(f_\eta) = \exp\left\{ j \frac{4\pi a_m}{c^2} \left[1 - \frac{D(f_\eta, u_{r_{ref}})}{D(f_{\eta_{ref}}, u_{r_{ref}})} \right] \left[\frac{R_0}{D(f_\eta, u_r)} - \frac{R_{ref}}{D(f_\eta, u_r)} \right]^2 \right\} \tag{5.62}$$

while the target phase restorer is of a simple form:

$$\hat{H}_{tp}(R_0) = \exp\left\{ j \frac{4\pi R_0}{c} \right\} \tag{5.63}$$

Finally, a focused image is obtained by the following operation:

$$S_5(\tau, \eta) = \left[\int_{-\infty}^{\infty} \left[S_4(\tau, f_\eta) \hat{H}_{az}(R_0, f_\eta) \cdot \hat{H}_{rp}(R_0, f_\eta) \right] \exp\{j 2\pi f_\eta \eta\} \, d\eta \right] \hat{H}_{tp}(R_0)$$

$$= A_2 p_r \left(\tau - \frac{2R_0}{cD(f_{\eta_{ref}}, u_{r_{ref}})} \right) g_a(\eta - \eta_c) \exp\left\{ j \frac{4\pi R_0}{c} \right\} \tag{5.64}$$

5.3.3 NUMERICAL EXAMPLES

In this section, we exam the property of the scaling function by two numerical cases: one uses a constant phase shift, while the other uses a linearly range-varying shift. Results after range compression with and without scaling are shown in Figures 5.11 and 5.12, respectively. Comparing with the original (unscaled) results, the scaling effect in the correcting phase error is clearly shown in both cases.

A comparison of focusing by CSA and RAD algorithms on a real data set provided in [12] is displayed in Figure 5.13. While no quantitative analysis is performed, we see that well-focused images are obtained by both CSA and RDA algorithms. Thus, the focusing procedures described above are validated.

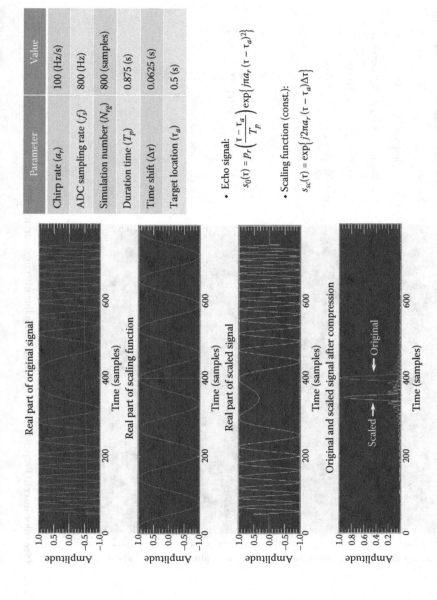

Parameter	Value
Chirp rate (a_r)	100 (Hz/s)
ADC sampling rate (f_s)	800 (Hz)
Simulation number (N_{rg})	800 (samples)
Duration time (T_p)	0.875 (s)
Time shift ($\Delta\tau$)	0.0625 (s)
Target location (τ_a)	0.5 (s)

- Echo signal:
$$s_0(\tau) = p_r\left(\frac{\tau - \tau_a}{T_p}\right)\exp\{j\pi a_r(\tau - \tau_a)^2\}$$

- Scaling function (const.):
$$s_{sc}(\tau) = \exp\{j2\pi a_r(\tau - \tau_a)\Delta\tau\}$$

FIGURE 5.11 **(See color insert.)** Property of scaling function with constant time shift after range compression.

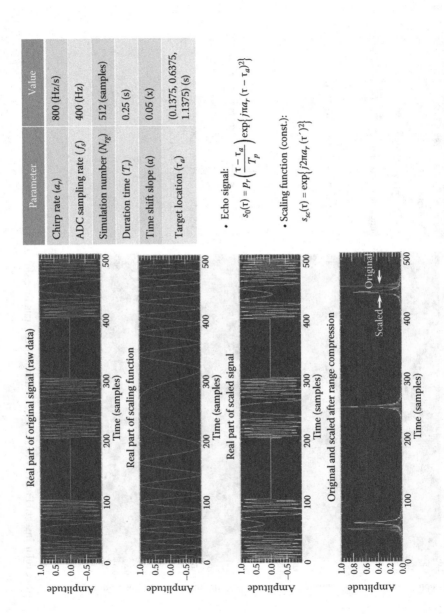

FIGURE 5.12 (See color insert.) Property of scaling function with linearly range-varying shift after range compression.

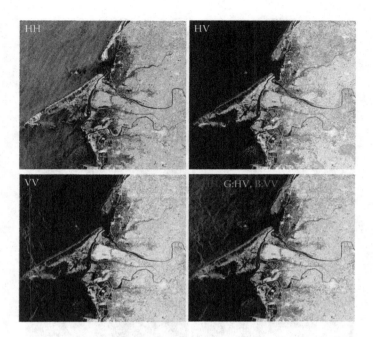

FIGURE 1.5 Polarization diversity enhances target feature (at Canada Space Agency [CSA]).

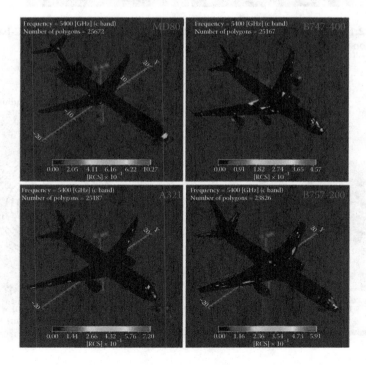

FIGURE 2.16 Computed RCS of commercial aircrafts for C-band HH Polarization (left) and X-band VV Polarization (right). *(Continued)*

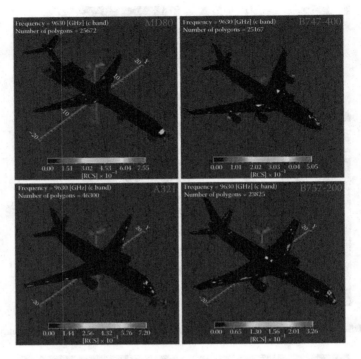

FIGURE 2.16 (CONTINUED) Computed RCS of commercial aircrafts for C-band HH Polarization (left) and X-band VV Polarization (right).

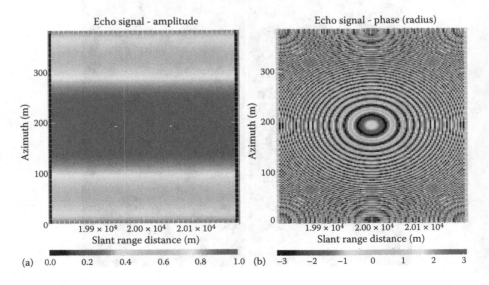

FIGURE 3.5 Typical echo signal after demodulation: amplitude (a) and phase (b).

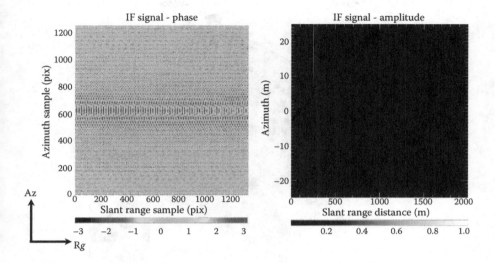

FIGURE 3.16 Phase and amplitude of the IF signal after Fourier transform on the range using the parameters in Table 3.1.

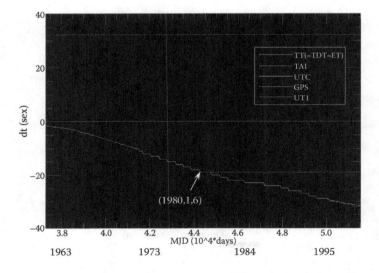

FIGURE 4.1 Time differences of TT, TAI, UTC, GPS Time, and UT1 (from top to bottom).

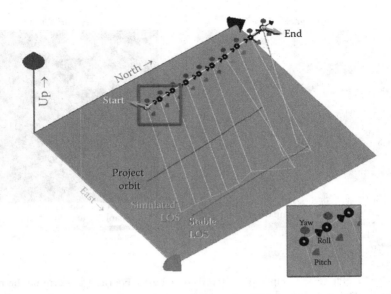

FIGURE 4.14 SAR slant range variations due to noisy flight path.

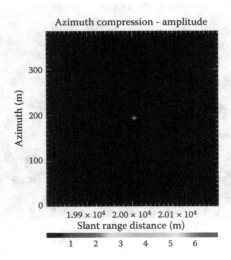

FIGURE 5.7 Focused image (amplitude) after azimuth compression.

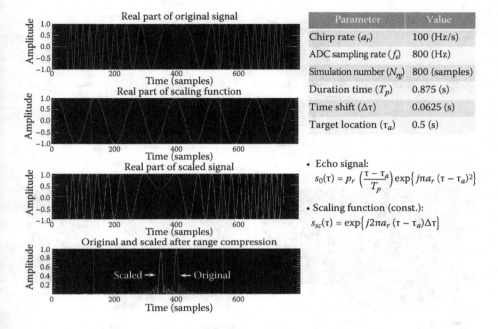

Parameter	Value
Chirp rate (a_r)	100 (Hz/s)
ADC sampling rate (f_s)	800 (Hz)
Simulation number (N_{rg})	800 (samples)
Duration time (T_p)	0.875 (s)
Time shift ($\Delta\tau$)	0.0625 (s)
Target location (τ_a)	0.5 (s)

- Echo signal:
$$s_0(\tau) = p_r\left(\frac{\tau - \tau_a}{T_p}\right)\exp\left\{j\pi a_r(\tau - \tau_a)^2\right\}$$

- Scaling function (const.):
$$s_{sc}(\tau) = \exp\left\{j2\pi a_r(\tau - \tau_a)\Delta\tau\right\}$$

FIGURE 5.11 Property of scaling function with constant time shift after range compression.

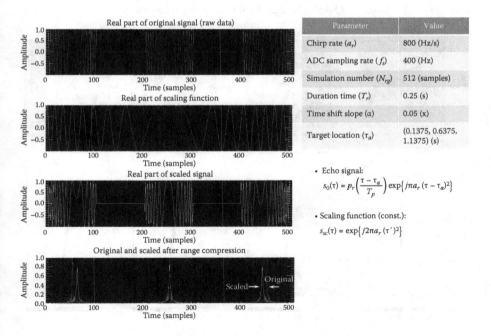

Parameter	Value
Chirp rate (a_r)	800 (Hz/s)
ADC sampling rate (f_s)	400 (Hz)
Simulation number (N_{rg})	512 (samples)
Duration time (T_r)	0.25 (s)
Time shift slope (α)	0.05 (x)
Target location (τ_a)	(0.1375, 0.6375, 1.1375) (s)

- Echo signal:
$$s_0(\tau) = p_r\left(\frac{\tau - \tau_a}{T_p}\right)\exp\left\{j\pi a_r(\tau - \tau_a)^2\right\}$$

- Scaling function (const.):
$$s_{sc}(\tau) = \exp\left\{j2\pi a_r(\tau')^2\right\}$$

FIGURE 5.12 Property of scaling function with linearly range-varying shift after range compression.

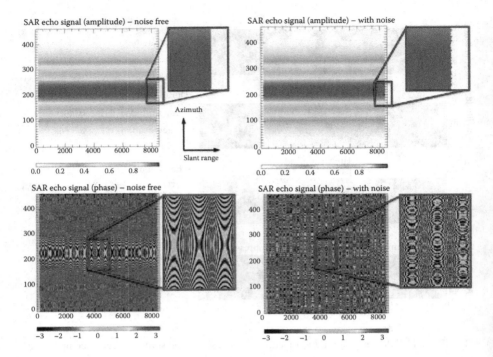

FIGURE 6.11 Simulated SAR echo signal with and without motion noise.

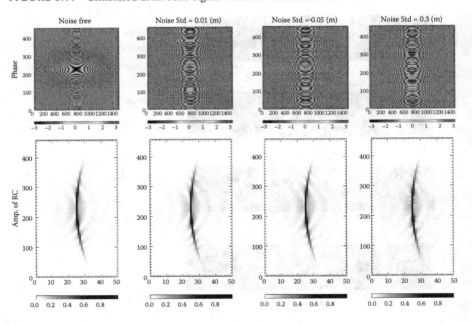

FIGURE 6.12 Simulated SAR echoes (amplitude and phase) with different noise standard deviation.

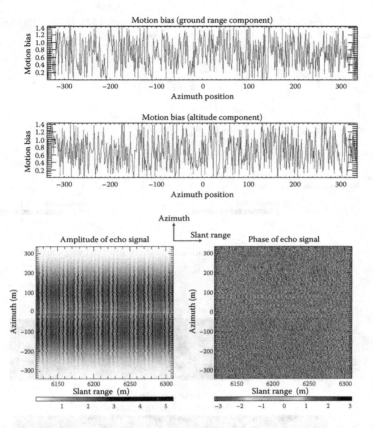

FIGURE 6.14 Simulated components of motion errors as the sensor moves (along the azimuth): bias in ground range (top) and bias in altitude (middle) and resulting echo signal (bottom).

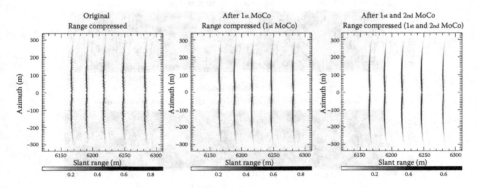

FIGURE 6.15 Illustration of first and second motion compensation effects. Data are displayed in the range compression–azimuth frequency domain.

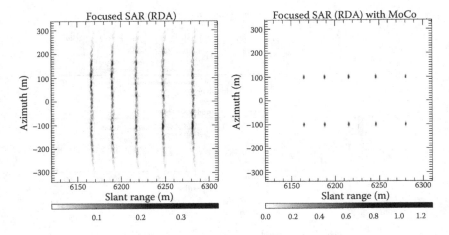

FIGURE 6.16 Focused image by RDA with and without motion compensation.

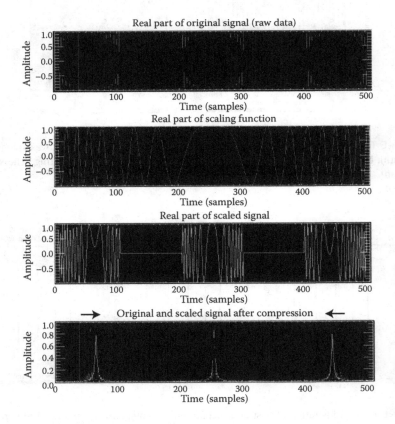

FIGURE 7.2 Basics of a chirp scaling operation.

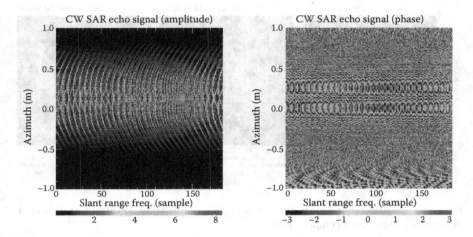

FIGURE 7.4 Echo signal of the imaging scenario given in Figure 7.3 with sensor parameters specified in Table 7.3.

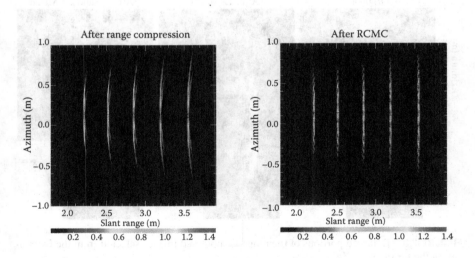

FIGURE 7.5 Simulated data after range compression and after RCMC.

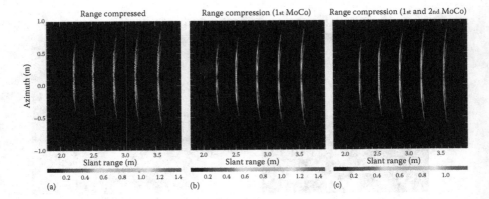

FIGURE 7.7 Effect of motion compensation, (a) no compensation, (b) after first compensation, and (c) after second compensation.

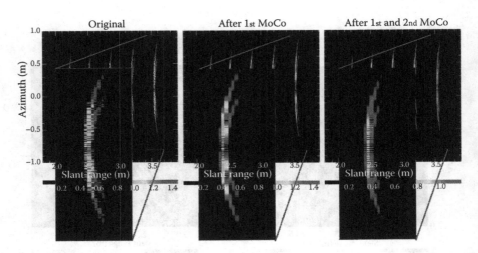

FIGURE 7.8 Enlarged portions of the range-compressed images selected from the two targets at the far range.

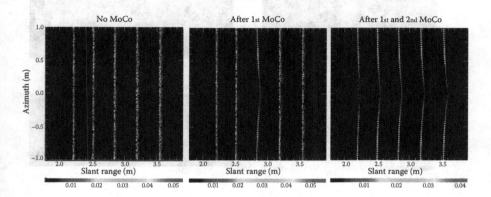

FIGURE 7.9 Illustration of first and second motion compensation to demonstrate the focusing effect.

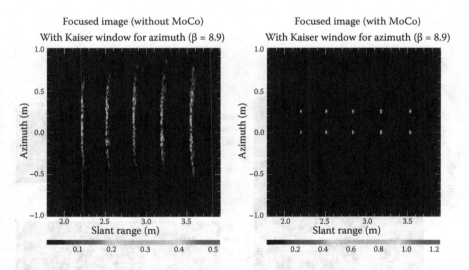

FIGURE 7.10 Focused image by RDA with and without motion compensation.

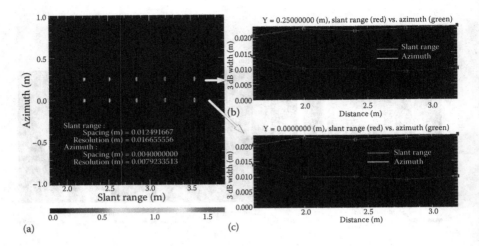

FIGURE 7.11 Geometric quality of the RDA focused image after MoCo. The upper right plot corresponds to the upper row of targets, and the lower right plot with the lower row of targets. (a) RDA focused image, (b) enlarged upper row target of (a), (c) enlarged upper row target of (a).

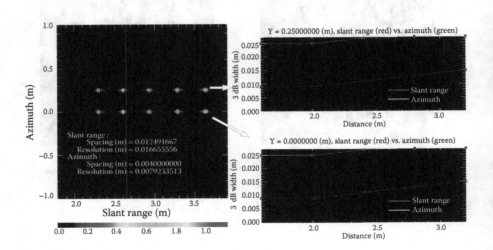

FIGURE 7.12 Similar to Figure 7.11, except focused by CSA.

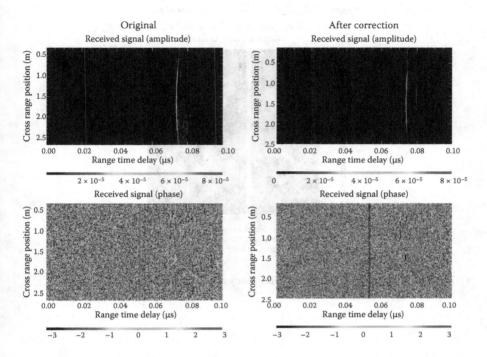

FIGURE 7.14 Original raw data (amplitude and phase) and corrected data.

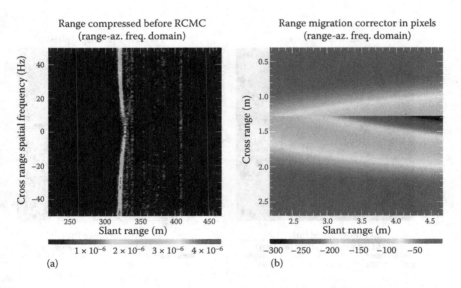

FIGURE 7.16 (a) Range migration revealed in the range–azimuth frequency domain and (b) required RCMC pattern.

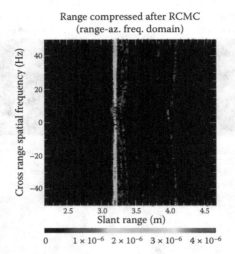

FIGURE 7.17 Data in range–azimuth frequency domain after RCMC.

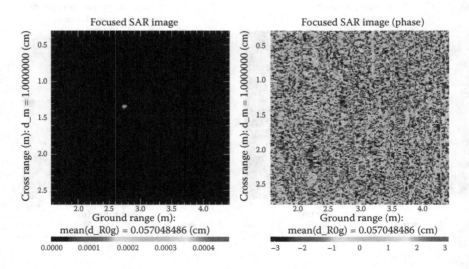

FIGURE 7.19 Focused image showing both amplitude and phase.

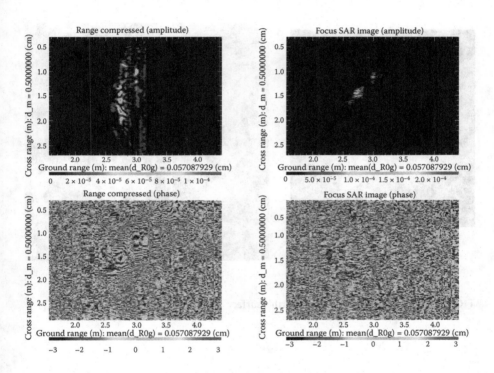

FIGURE 7.23 Range-compressed image and focused image of the target.

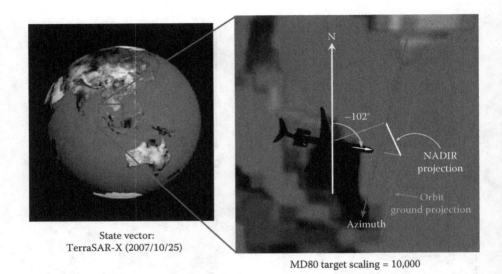

State vector:
TerraSAR-X (2007/10/25)

MD80 target scaling = 10,000

FIGURE 8.4 Target setting on the earth's surface for a TerraSAR image simulation.

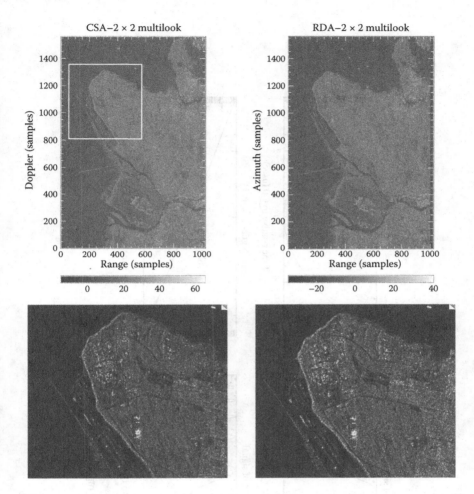

FIGURE 5.13   Comparison of focused images using RDA and CSA. The lower image clips are enlargements of the squared window portion, as indicated. SAR images copyright of CSA.

APPENDIX: STATIONARY PHASE APPROXIMATION

For easy reference, we give the principles of stationary phase approximation [23]. Figure A.1 is a typical chirp signal of the form in Equation 5.1 where the phase is rapidly time varying and the envelope is almost constant over one complete phase cycle. In the process of range compression and azimuth compression, Fourier transforms are performed. At this point, the contribution to the integral (both real and imaginary parts) is almost zero, because the positive and negative parts of the phase cycle cancel each other. Therefore, the contribution to the integral lies mainly around the stationary phase point. Below is the technique for the principles of the stationary phase.

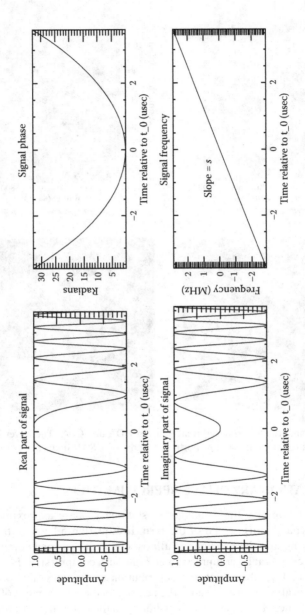

FIGURE A.1 Typical chirp signal indicating the rapid time-varying phase.

Consider the integral form of interest:

$$I = \int_a^b f(x)e^{j\lambda g(x)}\,dx \tag{A.1}$$

Assume that λ is a constant real number and

$$d^{(n)}g(x)/dx^{(n)} \to 0, \quad \exists n \to \infty \tag{A.2}$$

Furthermore, $f(x)$ is almost constant when $x \to x_s$.
The amplitude is approximated as

$$f(x) \approx f(x_s) \tag{A.3}$$

while the phase is expanded into Taylor series up to two terms:

$$g(x) \approx g(x_s) + \frac{g''(x_s)}{2!}(x - x_s)^2 \tag{A.4}$$

Then, the approximate to the integral in Equation A.1 becomes

$$
\begin{aligned}
I &= \int_a^b f(x)e^{j\lambda g(x)}\,dx \\[2mm]
&\approx f(x_s)e^{j\lambda g(x_s)} \int_a^b e^{j\lambda \frac{g''(x_s)}{2!}(x-x_s)^2}\,dx \\[2mm]
&\approx f(x_s)e^{j\lambda g(x_s)} \int_{-\infty}^{\infty} e^{j\lambda \frac{g''(x_s)}{2!}(x-x_s)^2}\,dx \\[2mm]
&= f(x_s)e^{j\lambda g(x_s)} \int_{-\infty}^{\infty} e^{j\lambda \frac{g''(x_s)}{2!}y^2}\,dy \\[2mm]
&= f(x_s)\sqrt{\frac{\pi}{\left|\lambda \dfrac{g''(x_s)}{2!}\right|}}\,\exp\left\{ j\left[\lambda g(x_s) + sign\left(\lambda \frac{g''(x_s)}{2!}\right)\frac{\pi}{4}\right]\right\}
\end{aligned}
\tag{A.5}
$$

If the phase is of the form

$$g(t) = w(t)\exp\{j\phi(t)\} \tag{A.6}$$

its Fourier transform is

$$G(f) = \int_{-\infty}^{\infty} g(t)\exp\{-j2\pi ft\}\,dt$$

$$= \int_{-\infty}^{\infty} w(t)\exp\{j\phi(t) - j2\pi ft\}\,dt \tag{A.7}$$

$$= \int_{-\infty}^{\infty} w(t)\exp\{j\theta(t)\}\,dt$$

The time-varying phase $\theta(t)$ is stationary at t_s if $\left.\dfrac{d\theta(t)}{dt}\right|_{t=t_s} = 0$. In such a case, the following approximation holds:

$$G(f) \approx CW(f)\exp\left\{j\left(\Theta(f) \pm \frac{\pi}{4}\right)\right\} \tag{A.8}$$

where

$$W(f) = w\{t(f)\}$$

$$\pm\frac{\pi}{4}\Theta(f) = \theta\{t(f)\}$$

$$C = \sqrt{\frac{2\pi}{\left|\dfrac{d^2\phi(t)}{dt^2}\right|_{t=t_s}}}$$

If C is constant, the constant phase, $\pm\dfrac{\pi}{4}$ in Equation A.8, may be ignored.

REFERENCES

1. Bamler, R., Doppler frequency estimation and the Cramer-Rao bound, *IEEE Transactions on Geoscience and Remote Sensing*, 25(3): 385–390, 1991.
2. Bamler, R., and Runge, H., PRF-ambiguity resolving by wavelength diversity, *IEEE Transactions on Geoscience and Remote Sensing*, 29(6): 997–1003, 1991.

3. Cumming, I. G., and Li, S., Improved slope estimation for SAR Doppler ambiguity resolution, *IEEE Transactions on Geoscience and Remote Sensing*, 44(3): 707–718, 2006.

4. Cumming, I. G., A spatially selective approach to Doppler estimation for frame-based satellite SAR processing, *IEEE Transactions on Geoscience and Remote Sensing*, 42(6): 1135–1148, 2004.

5. Jin, M. Y., Optimal Doppler centroid estimation for SAR data from a quasihomogeneous source, *IEEE Transactions on Geoscience and Remote Sensing*, 24: 1022–1025, 1986.

6. Kong, Y. K., Cho, B.-L., and Kim, Y.-S., Ambiguity-free Doppler centroid estimation technique for airborne SAR using the radon transform, *IEEE Transactions on Geoscience and Remote Sensing*, 43(4): 715–721, 2005.

7. Madsen, S. N., Estimating the Doppler centroid of SAR data, *IEEE Transactions on Aerospace and Electronic System*, 25: 134–140, 1989.

8. Prati, C., and Rocca, F., Focusing SAR data with time-varying Doppler centroid, *IEEE Transactions on Geoscience and Remote Sensing*, 30: 550–559, 1992.

9. Wong, F., and Cumming, I. G., A combined SAR Doppler centroid estimation scheme based upon signal phase, *IEEE Transactions on Geoscience and Remote Sensing*, 34(3): 696–707, 1996.

10. Cafforio, C., Prati, C., and Rocca, E., SAR data focusing using seismic migration techniques, *IEEE Transactions on Aerospace and Electronic Systems*, 27: 194–207, 1991.

11. Carrara, W. G., Majewski, R. M., and Goodman, R. S., *Spotlight Synthetic Aperture Radar: Signal Processing Algorithms*, Artech House, Norwood, MA, 1995.

12. Cumming, I. G., and Wong, F. H., *Digital Processing of Synthetic Aperture Radar Data: Algorithms and Implementation*, Artech House, Norwood, MA, 2005.

13. Curlander, J. C., and McDonough, R. N., *Synthetic Aperture Radar: Systems and Signal Processing*, Wiley-Interscience, New York, 1991.

14. Franceschitti, G., and Lanari, R., *Synthetic Aperture Radar Processing*, CRC Press, New York, 1999.

15. Richards, M. A., *Fundamentals of Radar Signal Processing*, 2nd ed., McGraw-Hill, New York, 2014.

16. Soumekh, M., *Synthetic Aperture Radar Processing*, John Wiley & Sons, New York, 1999.

17. Wahl, D. E., Eiche, P. H., Ghiglia, D. C., Thompson, P. A., and Jakowatz, C. V., *Spotlight-Mode Synthetic Aperture Radar: A Signal Processing Approach*, Springer, New York, 1996.

18. Gen, S. M., and Huang, F. K., A modified range-Doppler algorithm for de-chirped FM-CW SAR with a squint angle, *Modern Radar*, 29(11): 49–52, 2007.

19. Papoulis, A., *Fourier Integral and Its Applications*, McGraw-Hill, New York, 1962.

20. Moreira, A., and Huang, Y., Airborne SAR processing of highly squinted data using a chirp scaling approach with integrated motion compensation, *IEEE Transactions on Geoscience and Remote Sensing*, 32(5): 1029–1040, 1994.

21. Moreira, A., Mittermayer, J., and Scheiber, R., Extended chirp scaling algorithm for air- and spaceborne SAR data processing in Stripmap and ScanSAR imaging modes, *IEEE Transactions on Geoscience and Remote Sensing*, 34(5): 1123–1136, 1996.

22. Raney, R. K., Runge, H., Bamler, R., Cumming, I. G., and Wong, F. H., Precision SAR processing using chirp scaling, *IEEE Transactions on Geoscience and Remote Sensing*, 32: 786–799, 1994.

23. Strang, G., *Introduction to Applied Mathematics*, Cambridge Press, Wellesley, 1986.

6 Motion Compensation

6.1 INTRODUCTION

A synthetic aperture radar (SAR) sensor, by moving, is an essential element to create relative motion and thus to acquire the Doppler frequency and its rate, from which the target's direction is determined. Together with ranging, a target's position may be determined. Image focusing, presented in Chapter 5, is assumed with constant velocity and linear trajectory. In reality, nonlinear sensor motion results cause an image to be defocused if not appropriately accounted for in image focusing [1–9]. During the course of moving along the synthetic aperture path, the sensor motion also induces possible antenna pointing errors and introduces undesired phase variations, and subsequently causes image defocusing and blurred [7,8]. In Chapter 4, we described the SAR slant range variations due to a random and unstable flight path. Just like the coherent processing that inherently produces the speckle, the unwanted random motion–induced Doppler is embedded in the total Doppler bandwidth and is persistently unavoidable. Sometimes, depending on the sensor platform motion stability, the Doppler error is so profound that its compensation is more difficult than we thought. Hence, without compensating such an phase error in the focusing process, poor image quality is mostly likely expected. The defocusing by sensor motion can be due to the variation of tangent velocity and variation of instantaneous slant range. Because of the motion, the radial velocity between the sensor and the target being imaged also varies. Principally, the tangent velocity variation is compensated by changing the pulse repetition frequency (PRF), while the phase variations due to the line of sight (LOS) path difference between the actual and the reference path are corrected with Global Positioning System (GPS), inertial navigation system (INS), and inertial measurement units (IMU) measurements, followed by using the Kalman filter, if necessary, to smooth the position, velocity bias, and error from INS or IMU. The phase variations due to an unstable slant range may be corrected with position measured by INS and attitude measured by IMU, if the data sampling rate is sufficiently large. This type of compensation is termed data-based motion compensation (DBMC) [4,10–18]. Another commonly used technique is the so-called signal-based motion compensation (SBMC) [7,8,18–21]. In SBMC, three steps are devised: Doppler tracking and range tracking, and their compensation and correction, and autofocusing. One example is the phase gradient algorithm (PGA), which makes use of the phase increment in the azimuth direction [19,22,23] and corrects all range bins to the same Doppler (azimuth) frequency. In the range–Doppler algorithm, DBMC is carried out in raw data once it is acquired, while the second motion compensation, if desired, is executed after the range compression. As for the SBMC, generally it is done after the azimuth compression. The SBMC is regarded as higher-order correction beyond the DBMC. This chapter is devoted to demonstrating the strategy and procedure of motion compensation in the SAR image focusing chain. Instead of presenting and comparing various methods of compensation, focus is placed on a selective compensation algorithm under the framework of the range–Doppler algorithm (RDA) and chirp scaling algorithm (CSA) focusing algorithms.

6.2 PARAMETERS AND THEIR CONVERSIONS

6.2.1 AIRBORNE SYSTEM

From Chapter 4, the sensor geometry is recalled in Figure 6.1. In the geocentric coordinate, let φ, ϑ, h denote the longitude, latitude, and height, respectively. In Figure 6.1, we also know that the sensor coordinate is $\hat{x}_e, \hat{y}_n, \hat{z}_u$, with corresponding sensor's altitude $\theta_p, \theta_r, \theta_y$.

The origin of the sensor coordinate is denoted as $(\varphi_0, \vartheta_0, h_0)$. In the ENU (east, north, up) coordinate, we may set $(x_e, y_n, z_u) = (0, 0, 0)$. Taking the sensor's altitude into account when transforming from the geocentric coordinate to the sensor's coordinate,

$$\mathbf{P_{enu}} = \begin{bmatrix} x_e \\ y_n \\ z_u \end{bmatrix} = \begin{bmatrix} (\varphi - \varphi_0)(E_b + h)\cos\vartheta \\ (\vartheta - \vartheta_0)(E_a + h) \\ h - h_0 \end{bmatrix} \tag{6.1}$$

where E_b, E_a are the semiaxes.

Defining a transform matrix with known sensor roll, pitch, yaw (RPY) angles, $\theta_p, \theta_r, \theta_y$,

$$\mathbf{M_1} = \mathbf{R_x}(\theta_r)\mathbf{R_y}(\theta_p)\mathbf{R_z}(\theta_y)$$

$$= \begin{bmatrix} \cos\theta_p\cos\theta_y & -\cos\theta_p\sin\theta_y & \sin\theta_p \\ \sin\theta_r\sin\theta_p\cos\theta_y + \cos\theta_r\sin\theta_y & -\sin\theta_r\sin\theta_p\sin\theta_y + \cos\theta_r\cos\theta_y & -\sin\theta_r\cos\theta_p \\ -\cos\theta_r\sin\theta_p\cos\theta_y + \sin\theta_r\sin\theta_y & \cos\theta_r\sin\theta_p\sin\theta_y + \sin\theta_r\cos\theta_y & \cos\theta_r\cos\theta_p \end{bmatrix}$$

$$\tag{6.2}$$

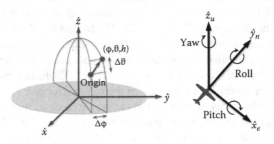

FIGURE 6.1   Airborne sensor motion geometry.

where the rotation matrices $R_x(\Theta)$, $R_y(\Theta)$, $R_z(\Theta)$ were given by Equation 4.3. Now transformation from the geocentric coordinate to the sensor's rectangular coordinate (x, y, z) is given by [24]

$$P_P = [x_p, y_p, z_p]^T = M_1 P_{enu} \tag{6.3}$$

With the given SAR look angle and squint angle, we may define the transformation matrix:

$$M_2 = R_y(-\theta_\ell)R_z(-\theta_{sq}) = \begin{bmatrix} \cos\theta_\ell & 0 & -\sin\theta_\ell \\ 0 & 1 & 0 \\ \sin\theta_\ell & 0 & \cos\theta_\ell \end{bmatrix} \begin{bmatrix} \cos\theta_q & \sin\theta_q & 0 \\ -\sin\theta_q & \cos\theta_q & 0 \\ 0 & 0 & 1 \end{bmatrix}$$

$$\tag{6.4}$$

Taking these two angles into account, it is convenient to use the LVP coordinate (Figure 6.2), which is obtained by the conversion of Equation 6.3:

$$P_d = [d_{los}, d_\parallel, d_\perp]^T = M_2 P_p = M_2 M_1 P_{enu} \tag{6.5}$$

For SAR processing, we need to simultaneously consider the sensor's altitude and the SAR looking geometry, which involves the matrices M_1 and M_2, respectively. The necessary matrix takes the form

$$M_3 = R_x(\theta_r)R_y(\theta_p - \theta_\ell)R_z(\theta_y - \theta_{sq}) \tag{6.6}$$

It should be noted that $M_3 \neq M_2 M_1$. If we further consider the looking direction, namely, either right looking or left looking, referred to as the sensor's heading, then we may write Equation 6.5 as

$$P_d = [d_{los}, d_\parallel, d_\perp]^T = M_3 M_0 P_{enu} \tag{6.7}$$

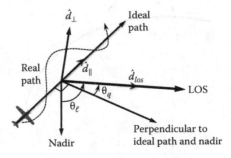

FIGURE 6.2 Schematic description of an aircraft motion in the LPV coordinate when taking the look angle and squint angle into account.

where

$$\mathbf{M_0} = \mathbf{R_z}(\theta_{RL}), \begin{cases} \theta_{RL} = +90°, \text{ for right looking} \\ \theta_{RL} = -90°, \text{ for left looking} \end{cases} \tag{6.8}$$

6.2.2 SPACEBORNE SYSTEM

For a satellite platform, the initial coordinate from the sensor to the LOS is considered in this section.

In SAR observation with a squint angle, the azimuth angle is defined by (Figure 6.3)

$$\theta_{az} = \tan^{-1}\left[\frac{R_0}{\sqrt{R_0^2 - (h+d)^2}} \tan(\theta_q)\right] \tag{6.9}$$

where

$$R_0 = \frac{1}{2}A \pm \sqrt{A^2 - 4[(R_e + h)^2 - R_e]} \tag{6.10}$$

$$A = 2(R_e + h)\cos(\theta_\ell), \quad R_0 > 0 \tag{6.11}$$

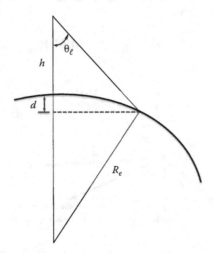

FIGURE 6.3 Simple SAR geometry.

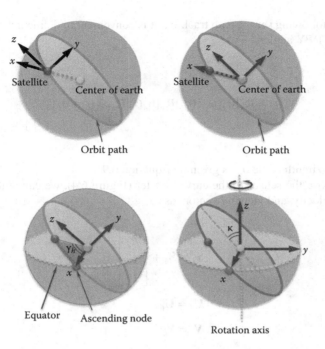

FIGURE 6.4 SAR sensors coordinate transformations.

Referring to Figure 6.4, suppose that the initial sensor position and velocity are expressed as a function of time t:

$$\mathbf{S}_0(t) = \begin{bmatrix} \Delta x \\ \Delta y \\ \Delta z \end{bmatrix} + \mathbf{U}_1 t + \mathbf{S}_0(t=0) \tag{6.12}$$

$$\mathbf{U}_0(t) = \begin{bmatrix} \Delta u_x \\ \Delta u_y \\ \Delta u_z \end{bmatrix} + \mathbf{U}_0(t=0) \tag{6.13}$$

In the above expressions, we note that the position and velocity vectors may be written as a sum of the nominal and bias or error terms, Δx, Δy, Δz and Δu_x, Δu_y, Δu_z, respectively. The initial conditions are simply given by

$$\mathbf{S}_0 = [0,0,0]^{\mathrm{T}}$$
$$\mathbf{U}_0 = [0,0,u]^{\mathrm{T}} \tag{6.14}$$

where u has been denoted as SAR speed in Chapter 4.

For the following sensor path tracking, it is convenient to define a viewing vector in terms of RPY angles:

$$\mathbf{V_0}(t) = \mathbf{R_x}[\theta_y(t) - \theta_{az}]\mathbf{R_y}[\theta_p(t)]\mathbf{R_z}[\theta_r(t) + \theta_\ell]\begin{bmatrix} 0 \\ 1 \\ 0 \end{bmatrix} \qquad (6.15)$$

where the azimuth angle θ_{az} is given by Equation 6.9.

If we move the sensor to the earth's center (Figure 6.4), we can denote the new position, velocity, and viewing vectors as

$$\mathbf{S_1} = \mathbf{S_0} + \begin{bmatrix} R_e \\ 0 \\ 0 \end{bmatrix}$$

$$\mathbf{U_1} = \mathbf{U_0}$$

$$\mathbf{V_1} = \mathbf{V_0} \qquad (6.16)$$

Now, let's point the x-axis to the ascending node or descending node; we can express the new position, velocity, and viewing vectors as

$$\mathbf{S_2} = \mathbf{T_{12}}\mathbf{S_1}$$

$$\mathbf{U_2} = \mathbf{T_{12}}\mathbf{U_1}$$

$$\mathbf{V_2} = \mathbf{T_{12}}\mathbf{U_1} \qquad (6.17)$$

where the transformation matrix is

$$\mathbf{T_{12}} = \mathbf{R_y}(-\gamma_h) = \begin{bmatrix} \cos\gamma_h & 0 & -\sin\gamma_h \\ 0 & 1 & 0 \\ \sin\gamma_h & 0 & \cos\gamma_h \end{bmatrix} \qquad (6.18)$$

with the hour angle γ_h determined by

$$\gamma_h = \omega_s t_s \qquad (6.19)$$

In Equation 6.19, note that from Figure 6.4, the hour angle is measured from the ascending node crossing; ω_s is the satellite angular velocity and t_s is the time since the ascending node crossing.

The next step is to transform the position, velocity, and viewing vectors to the earth's center. From Figure 6.4, with the angle between the z-axis and the plane of the SAR orbit path denoted as γ, the transformation matrix we need is

$$\mathbf{T}_{23} = R_x(\kappa) = \begin{bmatrix} 1 & 0 & 0 \\ 0 & \cos\kappa & -\sin\kappa \\ 0 & \sin\kappa & \cos\kappa \end{bmatrix} \tag{6.20}$$

Then the resulting vectors are obtained by

$$\mathbf{S}_3 = \mathbf{T}_{23}\mathbf{S}_2$$
$$\mathbf{U}_3 = \mathbf{T}_{23}\mathbf{U}_2 \tag{6.21}$$
$$\mathbf{V}_3 = \mathbf{T}_{23}\mathbf{U}_2$$

At this point, let's consider the relative position and velocity vectors between the sensor and target on the earth's surface in order to calculate the Doppler frequency. With the sensor position vector given in Equation 6.21, the target position vector can be easily derived as (Figure 6.5)

$$\mathbf{P}_3 = \mathbf{S}_3 + R_3\mathbf{V}_{3g} \tag{6.22}$$

where R_3 is the slant range to be determined and the viewing matrix $\mathbf{V}_{3g}$ is given by

$$\mathbf{V}_{3g} = \mathbf{T}_{z3}^{-1}\mathbf{T}_{y3}\mathbf{T}_{z3}\mathbf{V}_3 \tag{6.23}$$

with

$$\mathbf{T}_{y3} = \mathbf{R}_y(\vartheta_g) \tag{6.24}$$

$$\mathbf{T}_{z3} = \mathbf{R}_z(\varphi_{sat}) \tag{6.25}$$

In Equation 6.25, ϑ_g is the difference between the geodetic and geocentric latitudes; φ_{sat} is the sensor's longitude.

To find the slant range, we make use of the earth's ellipsoid equation:

$$\frac{P_{3x}^2}{E_a^2} + \frac{P_{3y}^2}{E_a^2} + \frac{P_{3z}^2}{E_b^2} = 1 \tag{6.26}$$

Notice that $\mathbf{P}_3 = [P_{3x}, P_{3y}, P_{3z}]^{\mathrm{T}}$.

The next step is to determine the target velocity. Suppose that the target moves with respect to the earth's surface (earth central rotational [ECR] coordinate). For

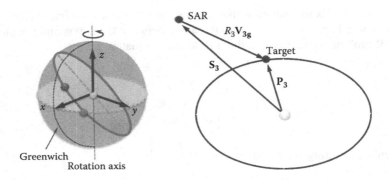

FIGURE 6.5 Determination of relative position vector between SAR and target.

simplicity without loss of generality, assume that only the y-component exists, that is, $\mathbf{U}_{tar} = [0\ D_3\omega_e\ 0]^T$, where ω_e is the earth's angular velocity and from Figure 6.6, $D_3 = \sqrt{P_{3y}^2 + P_{3x}^2}$.

The target velocity vector with respect to the earth central inertial (ECI) frame is

$$\mathbf{Q}_3 = \mathbf{T}_{43}\mathbf{U}_{tar} \tag{6.27}$$

where the transformation matrix is

$$\mathbf{T}_{43} = \mathbf{R}_z(\varphi_{tar}) \tag{6.28}$$

with φ_{tar} being the target's longitude. Now the relative target velocity (radial velocity) can be obtained from the difference between the sensor velocity and the target velocity:

$$u_r = \mathbf{U}_3 \cdot \mathbf{V}_{3g} - \mathbf{Q}_3 \cdot \mathbf{V}_{3g} = (\mathbf{U}_3 - \mathbf{Q}_3) \cdot \mathbf{V}_{3g} \tag{6.29}$$

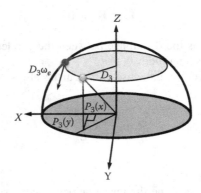

FIGURE 6.6 Geometry of target velocity with respect to ECR.

Notice that if the target is moving instead of stationary, then its velocity should be accounted for in a straightforward manner.

6.3 SIGNAL MODEL FOR MOTION COMPENSATION

In this section, we consider the signal model of motion compensation (MoCo) for the pulse system. Moreira and coworkers [14–16] proposed an extended chirp scaling algorithm for airborne and spaceborne SAR data processing, including the two-step motion compensations. The processing flowchart is illustrated in Figure 6.7, in which several key features deserve a closer look. The first MoCo is performed in the time domain before doing the azimuth fast Fourier transform (FFT), followed by chirp scaling for differential range cell migration correction (RCMC), which is in the range–Doppler domain. After the inverse FFT in azimuth, the second MoCo is carried out, again in the time domain. The SBMC step may be applied for final refinement, if necessary. It is readily noticed that motion compensations are all performed in the time domain.

Recalling that an echo signal returned from the target with range dependence $R_{\text{real}}(\eta)$,

$$s_{if}(\tau, \eta) = g_r(\tau) g_a(\eta) \exp\left\{ j2\pi \left[f_c \frac{2R_{\text{real}}(\eta)}{c} + a_r t \frac{2R_{\text{real}}(\eta)}{c} - \frac{2a_r}{c^2} R_{\text{real}}^2(\eta) \right] \right\}$$

(6.30)

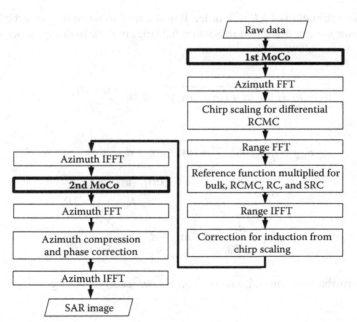

FIGURE 6.7 Processing flowchart of motion compensation under the framework of the extended chirp scaling algorithm [16].

let's define the actual slant range as a sum of an ideal (unperturbed) term and an error term, which is caused by the sensor motion:

$$R_{real}(\eta) = R(\eta) + \Delta R(\eta) \tag{6.31}$$

Then Equation 6.30 is reexpressed as follows, after the slant range is replaced by Equation 6.31:

$$s_{if}(\tau, \eta) = g_r(\tau) g_a(\eta) \exp\left\{ j2\pi \left[f_c \frac{2R(\eta)}{c} + a_r\tau \frac{2R(\eta)}{c} - \frac{2a_r}{c^2} R(\eta)^2 \right] \right\}$$

$$\times \exp\left\{ -j2\pi \left[f_c \frac{2\Delta R(\eta)}{c} + a_r\tau \frac{2\Delta R(\eta)}{c} - \frac{4a_r}{c^2} R(\eta)\Delta R(\eta) + \frac{2a_r}{c} \Delta R(\eta)^2 \right] \right\} \tag{6.32}$$

The phase associated with the perturbation term ΔR to be corrected is

$$\phi_{MC1} = \exp\left\{ -j2\pi \left[f_c \frac{2\Delta R(\eta)}{c} + a_r t \frac{2\Delta R(\eta)}{c} - \frac{4a_r}{c^2} R(\eta)\Delta R(\eta) + \frac{2a_r}{c^2} \Delta R(\eta)^2 \right] \right\} \tag{6.33}$$

Now the estimation of ΔR is in order. It is practical to relate the perturbation term to the sensor motion described in Section 6.2 (Figure 6.2). In doing so, let's expand Equation 6.3 as follows [16]:

$$R(\eta) = \sqrt{R_0^2 + (u\eta - u\eta_0)^2} \approx R_0 + \frac{(u\eta - u\eta_0)^2}{2R_0} \tag{6.34}$$

$$R_{real} = \sqrt{(R_0 - d_{los})^2 + (u\eta - u\eta_0 + d_{\parallel})^2 + d_\perp^2}$$

$$\approx R(\eta) + \frac{(u\eta - u\eta_0)^2}{2R_0} + \frac{(u\eta - u\eta_0)d_{\parallel}}{R_0} + \frac{d_{\parallel}^2}{2R_0} \tag{6.35}$$

$$- d_{los} + \frac{d_{los}}{2R_0^2} (u\eta - u\eta_0 + d_{\parallel})^2 + \frac{d_\perp^2}{2R_0} + \frac{d_{los}}{2R_0^2} d_\perp^2$$

The perturbation term in Equation 6.35 is now readily given by

$$\Delta R = d_{los} - \frac{d_{los}}{2R_0^2} (\eta - \eta_0 + d_{\parallel})^2 - \frac{d_\perp^2}{2R_0} - \frac{d_{los}}{2R_0^2} d_\perp^2 - \frac{(\eta - \eta_0)d_{\parallel}}{R_0} - \frac{d_{\parallel}^2}{2R_0} \tag{6.36}$$

Remember that the component $d_\parallel$, as defined in Figure 6.2, is the velocity component in the azimuth. When d_{los} and $\eta - \eta_0$ are both sufficiently small, $\eta - \eta_0$ may be replaced by η for faster computation. For low- to mediate-resolution systems, normally only the first term in Equation 6.36 is corrected [17,25–28].

In [18], the first-order MoCo (called bulk MoCo) is to correct to the reference range, R_{ref}, referred to as the imaging scene center. Hence, it is expected that correction performance will be degraded for those some distance away from the scene center, as will be shown in later simulation results. To compensate this defect, the second-order correction is devised [29–31].

The matched filter for first-order MoCo is of the form

$$H_{MC1} = \exp\left\{-j2\pi\left[f_c\frac{2R_{ref}(\eta)}{c} + a_r\tau\frac{2R_{ref}\eta}{c}\right]\right\} \qquad (6.37)$$

It is noted that for pulse systems, the first MoCo is performed in raw data domain, while for continuous-wave systems, it is done in the intermediate echo signal. The second term on the right-hand side of Equation 6.37 is generally ignored in pulse SAR because τ is very small.

The matched filter for the second order is mainly to correct the residual term, d_{los}, in Equation 6.36.

$$H_{MC2}(\eta, R_{ref}) = \exp\left\{-j2\pi\left[f_c\frac{2[\Delta R(R_0, \eta) - R_{ref}(\eta)]}{c}\right]\right\} \qquad (6.38)$$

Notice that the range displacement $\Delta R(R_0, \eta) - R_{ref}(\eta)$ is generic two-dimensional data.

6.4   NUMERICAL SIMULATION

In Chapter 4, we illustrated the procedures of simulating a sensor trajectory through multiple coordinate transformations that line up the target position, SAR, and platform into a uniform coordinate. This section adopts the same procedure to simulate the motion error for the purpose of motion compensation in image focusing. This provides a useful scheme to evaluate, for a SAR system specification, the focusing performance to achieve the desired image quality, including spatial resolution and contrast. Following the same simulation parameter for the SAR geometry specified in Table 4.1, the simulated motion errors are listed in Table 6.1. The measurable bias includes the sensor position (x, y, z), altitude (roll, pitch, yaw), and velocity, while the unmeasurable noise comes with absolute and relative terms. For each test run, the total number of simulation samples was 1751, based on the PRF we selected (600 Hz). Gaussian noise with zero mean was assumed. Figure 6.8 displays an instantaneous aircraft motion tunnel for the motion parameters given in Table 6.1. Figure 6.9 summarizes the complete flowchart of path trajectory simulation for SAR motion compensation in image focusing.

From Figure 6.9, we see that three generators are responsible for the bias, path trajectory, and noise; their inputs are described and given in Table 6.1. Note that the

TABLE 6.1
Simulation Motion Errors for an Airborne SAR System

Bias (Measurable)	
Parameter	Numeric Value
Positive bias (m, m, m)	3, 3, 3
Roll, pitch, and yaw bias (deg, deg, deg)	2, 2, 2
Velocity bias (m/s)	5

Noise (Unmeasured)		
Accuracy	Parameter	Numeric Value
Absolute	Position (m)	0.05 ~ 0.3
	Velocity (m/s)	0.005
	Roll angle (deg)	0.005
	Pitch angle (deg)	0.005
	True heading (yaw angle) (deg)	0.008
Relative	Noise (deg/sqrt(h))	<0.02
	Drift (deg/h)	0.1

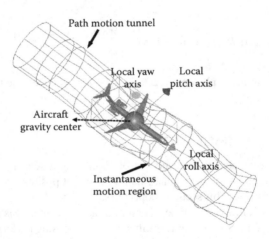

FIGURE 6.8   Aircraft motion tunnel.

position bias and noise are in the ECR coordinate. The noisy line-of-sight, perpendicular, velocity (LPV) parameters are generated through the SAR look angle and squint angle with given inputs from the altitude rotation in the RPY coordinate. The resulting motion errors (noisy trajectory) are then used to generate the state vector (SV) and feed it into SAR raw data generation. The echo signal is further processed into an image, with input bias-generated SV, to evaluate the motion compensation performance. The procedure is illustrated in Figure 6.10.

Figure 6.11 displays a simulated echo signal (both amplitude and phase) from a point target with and without motion error, assuming a zero squint angle. Irregular

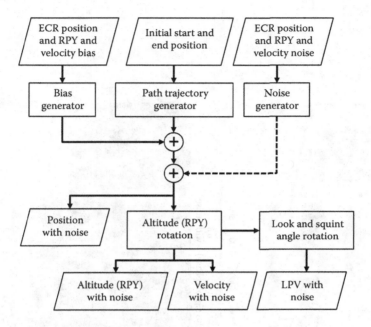

FIGURE 6.9 Simulation flowchart of the SAR path trajectory.

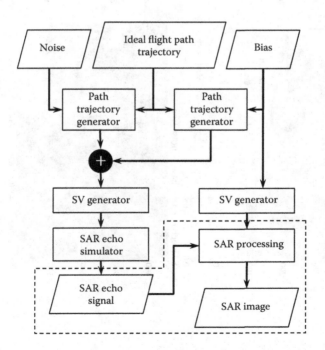

FIGURE 6.10 Flowchart of the flight trajectory for SAR echo signal simulation.

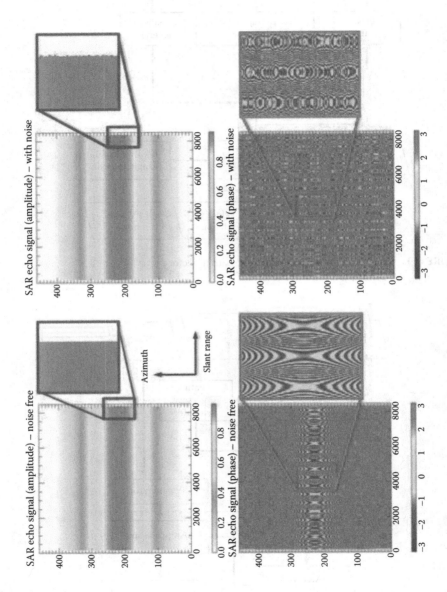

FIGURE 6.11   (See color insert.) Simulated SAR echo signal with and without motion noise.

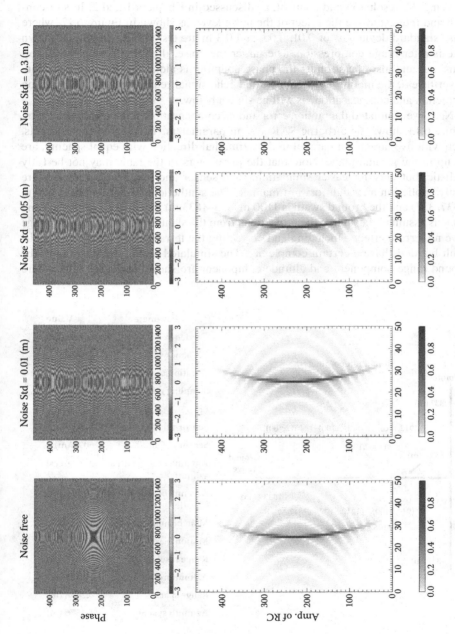

FIGURE 6.12 (See color insert.) Simulated SAR echoes (amplitude and phase) with different noise standard deviation.

amplitude along the azimuth direction is evident. The phase distortion is even more pronounced. Hence, both distorted amplitude and phase will blur the final image by lowering the resolution and contrast, as discussed in Chapters 1 and 2. It is of concern and interest to see the effect of the noise level, as shown in Figure 6.12, where noise standard deviations of 0.01, 0.05, and 0.3 m are tested. The amplitude shown here is after range compression. The larger the noise, the more severe the distortions are, and also the stronger the phase coupling is along the azimuth direction. Nevertheless, the motion-induced distorted echo signal requires appropriate motion correction and compensation, as will be shown below.

Next, we simulated the motion error and its compensation in the airborne system. Figure 6.13 shows the airborne SAR system parameters. Two sets of point targets, each with five targets, in parallel in the azimuthal direction with equal spacing, are set up in the ground plane. Note that the parameters in the table may not be fully realistic, but only for easier simulation. Real sensor and platform parameters are easily applied in a straightforward manner. The synthetic aperture length is 674 m (−337, 337) and the ground swath is 1100 m (533–1633 m). For simplicity, the ground plane is assumed flat so that the projections from the slant range to the ground range have no terrain effects. Including and correcting the terrain slope effects is not difficult but does have a certain complexity. The simulated motion biases in both the ground range component and altitude component are shown in Figure 6.14, where

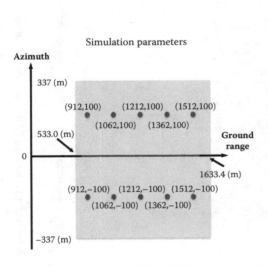

Parameter	Value
Carrier freq.	L: 1.4 (GHz)
TX bandwidth	170 (MHz)
Duration time	42 (μs)
Sampling rate	600 (MHz)
PRF	60 (Hz)
Sensor velocity	314.3 (km/h)
Sensor height	6096 (m)
Antenna beamwidth at azimuth	2 (degrees)
Antenna beamwidth at elevation	10 (degrees)
Main beam look angle	10 (degrees)
Motion bias std.	5 (m)
Range resolution	0.8818 (m)
Azimuth resolution	3.0671 (m)
Range spacing	0.2498 (m)
Azimuth spacing	1.4576 (m)

FIGURE 6.13 Simulation parameters and target locations set up in the azimuth–ground range plane.

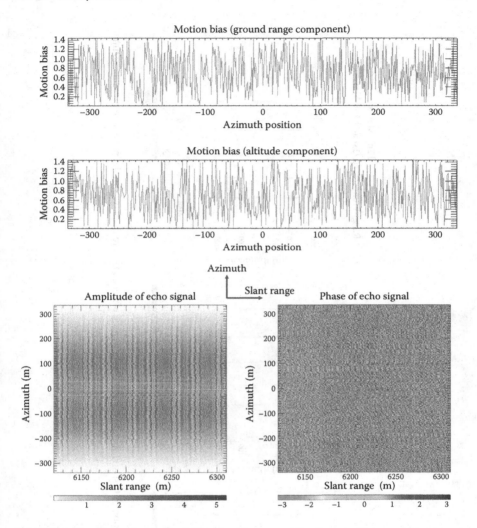

FIGURE 6.14   (See color insert.) Simulated components of motion errors as the sensor moves (along the azimuth): bias in ground range (top) and bias in altitude (middle) and resulting echo signal (bottom).

the resulting echo signals of the amplitude and phase are also shown. The properties of motion-induced distorted signals were explained previously. We discuss the motion compensation in the following.

Figure 6.15 illustrates the results of motion compensation after first and second MoCo. For better visual inspection, data are displayed in the range compression–azimuth frequency domain. The blurred effect along the azimuthal direction without MoCo is to show that only the first MoCo is not sufficient to correct the motion effect. After the second MoCo, a much better result is obtained, as shown here and

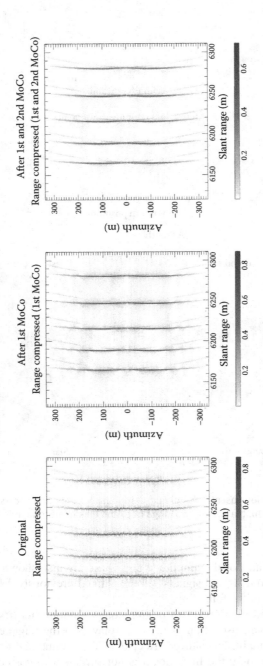

FIGURE 6.15 **(See color insert.)** Illustration of first and second motion compensation effects. Data are displayed in the range compression–azimuth frequency domain.

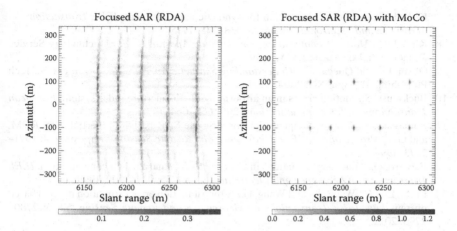

FIGURE 6.16 **(See color insert.)** Focused image by RDA with and without motion compensation.

later in the final focusing. As described earlier in this section, the second MoCo is done through a differential LOS correction right after the first MoCo, the bulk MoCo.

The final focused image by RDA with and without MoCo is shown in Figure 6.16. Without MoCo, the data are actually in no way able to focus into an image, which is totally unacceptable. Of course, this is already well known. By applying the procedure outlined in this section, one may be able to track the motion effects in each stage of image focusing under the framework of RDA and CSA. We will defer the quantitative check on image quality to Chapter 7, where the ground-based frequency-modulated continuous-wave (FMCW) system is presented.

REFERENCES

1. Fornaro, G., Franceschetti, G., and Perna, S., Motion compensation errors: Effects on the accuracy of airborne SAR images, *IEEE Transactions on Aerospace and Electronic Systems*, 41(4): 1338–1352, 2005.
2. Franceschitti, G., and Lanari, R., *Synthetic Aperture Radar Processing*, CRC Press, Boca Raton, FL, 1999.
3. Franceschetti, G., Iodice, A., Perna, S., and Riccio, D., SAR sensor trajectory deviations: Fourier domain formulation and extended scene simulation of raw data, *IEEE Transactions on Geoscience and Remote Sensing*, 44(9): 2323–2334, 2006.
4. Isernia, T., Pascazio, V., Pierri, R., and Schirinzi, R. R., Synthetic aperture radar imaging from phase-corrupted data, *IEE Proceedings—Radar, Sonar and Navigation*, 143(4): 268–274, 1996.
5. Khwaja, A. S., Ferro-Famil, L., and Pottier, E., Efficient SAR raw data generation for anisotropic urban scenes based on inverse processing, *IEEE Geoscience and Remote Sensing Letters*, 6(4): 757–761, 2009.
6. Khwaja, A. S., Ferro-Famil, L., and Pottier, E., Efficient Stripmap SAR raw data generation taking into account sensor trajectory deviations, *IEEE Geoscience Remote Sensing Letter*, 8(4): 794–798, 2011.

7. Kirk, J. C., Motion compensation for synthetic aperture radar, *IEEE Transactions on Aerospace and Electronic Systems*, 3: 338–348, 1975.

8. Kirk, J. C., *Motion Compensation for Synthetic Aperture Radar*, Technology Service Corporation, Los Angeles, CA, 1999.

9. Oliver, C., and Quegan, S., *Understanding Synthetic Aperture Radar Images*, SciTech Publishing, Raleigh, NC, 2004.

10. Buckreuss, S., Motion errors in an airborne synthetic aperture radar system, *European Transactions on Telecommunications*, 2(6): 655–664, 1991.

11. Buckreuss, S., Motion compensation for airborne SAR based on inertial data, RDM, and GPS, *Proceedings of IEEE Geoscience and Remote Sensing Symposium*, 4: 1971–1973, 1994.

12. Fornaro, G., Trajectory deviations in airborne SAR: Analysis and compensation, *IEEE Transactions on Aerospace and Electronic Systems*, 35: 997–1009, 1999.

13. Li, Y., Liu, C., Wang, Y., and Wang, Q., A robust motion error estimation method based on raw data, *IEEE Transactions on Geoscience and Remote Sensing*, 50(7): 2780–2790, 2012.

14. Moreira, J. R., A new method of aircraft motion error extraction from radar raw data for real time motion, *IEEE Transactions on Geoscience and Remote Sensing*, 28(4): 620–626, 1990.

15. Moreira, A., and Huang, Y., Airborne SAR processing of highly squinted data using a chirp scaling approach with integrated motion compensation, *IEEE Transactions on Geoscience and Remote Sensing*, 32(5): 1029–1040, 1994.

16. Moreira, A., Mittermayer, J., and Scheiber, R., Extended chirp scaling algorithm for air-and spaceborne SAR data processing in Stripmap and ScanSAR imaging modes, *IEEE Transactions on Geoscience and Remote Sensing*, 34(5): 1123–1136, 1996.

17. Reigber, A., Alivizatos, E., Potsis, A., and Moreira, A., Extended wavenumber-domain synthetic aperture radar focusing with integrated motion compensation, *IEEE Proceedings—Radar, Sonar and Navigation*, 153(3): 301–310, 2006.

18. Zaugg, E. C., and Long, D. G., Theory and application of motion compensation for LFM-CW SAR, *IEEE Transactions on Geoscience and Remote Sensing*, 46: 2990–2998, 2008.

19. Chan, H. L., and Yeo, T. S., Noniterative quality phase gradient autofocus (QPGA) algorithm for Spotlight SAR imagery, *IEEE Transactions on Geoscience and Remote Sensing*, 36: 1531–1539, 1998.

20. De Macedo, K. A. C., and Scheiber, R., Precise topography- and aperture-dependent motion compensation for airborne SAR, *IEEE Geoscience and Remote Sensing Letters*, 2(2): 172–176, 2005.

21. De Macedo, K. A. C., Scheiber, R., and Moreira, A., An autofocus approach for residual motion errors with application to airborne repeat-pass SAR interferometry, *IEEE Transactions on Geoscience and Remote Sensing*, 46(10): 3151–3162, 2008.

22. Van Rossum, W. L., and Otten, M., Extended PGA for range migration algorithms, *IEEE Transactions on Aerospace and Electronic Systems*, 42: 478–488, 2006.

23. Wahl, D., Eichel, P. H., Ghiglia, D. C., and Jakowatz, C. V., Phase gradient autofocus: A robust tool for high resolution SAR phase correction, *IEEE Transactions on Aerospace and Electronic Systems*, 30(3): 827–835, 1994.

24. Zhang, M., Liu, C., and Wang, Y. F., Motion compensation for airborne SAR with synthetic bandwidth, *Journal of Electronics and Information Technology*, 33(9): 2114–2119, 2011.

25. Rodriguez-Cassola, M., Prats, P., Krieger, G., and Moreira, A., Efficient time-domain image formation with precise topography accommodation for general bistatic SAR configurations, *IEEE Transactions on Aerospace and Electronic Systems*, 47(4): 2949–2966, 2011.

26. Stevens, D. R., Cumming, I. G., and Gray, A. L., Options for airborne interferometric SAR motion compensation, *IEEE Transactions on Geoscience and Remote Sensing*, 33: 409–419, 1995.
27. Sun, G., Jiang, X., Xing, M., and Qiao, Z.-J., Focus improvement of highly squinted data based on azimuth nonlinear scaling, *IEEE Transactions on Geoscience and Remote Sensing*, 49(6): 2308–2322, 2011.
28. Vandewal, M., Speck, R., and Süß, H., Efficient SAR raw data generation including low squint angles and platform instabilities, *IEEE Geoscience Remote Sensing Letter*, 5(1): 26–30, 2008.
29. Wu, H., and Zwick, T., Micro-air-vehicle-borne near-range SAR with motion compensation, *Progress in Electromagnetics Research*, 145: 11–18, 2014.
30. Feng, Y. C., Motion compensation in airborne synthetic aperture radar signal, Thesis, Institute of Mechanical Engineering, College of Engineering National Chiao Tung University, 2004.
31. Ozan, D., and Kartal, M., Efficient Stripmap-mode SAR raw data simulation including platform angular deviations, *IEEE Geoscience Remote Sensing Letter*, 8(4): 784–788, 2011.

7 Stationary FMCW SAR

7.1 INTRODUCTION

This chapter deals with a stationary frequency-modulated continuous-wave (SFMCW) system for which the synthetic aperture radar (SAR) sensor is quasi-stationary. Unlike the conventional SAR that makes use of the Doppler frequency by sensor motion, the SFMCW SAR, with no Doppler information available, must have a new signal model to design a proper focusing algorithm. In light of this, in this chapter, we derive a mathematical signal model, followed by developing a focusing algorithm based on the range–Doppler and chirp scaling algorithms, which were generally outlined in Chapter 5. Numerical simulations and indoor and outdoor experiments were designed and validated the system. Also presented is the motion compensation (MoCo), a sequel to Chapter 6.

7.2 SFMCW SIGNAL MODEL

Both the pulse and frequency-modulated continuous-wave SAR system use Doppler information to achieve fine resolution in the azimuthal direction. This is not possible for a stationary or quasi-stationary SAR system. A pseudo-Doppler must be synthetized. A linear FMCW signal is of the form [1–3]

$$s_t(\tau) = \exp\left\{ j2\pi\left(f_c\tau + \frac{1}{2}a_r\tau^2 \right) \right\} \tag{7.1}$$

where f_c is the carrier frequency and a_r is chirp rate. Assuming a point target model for simplicity, the received signal in the slant direction with a delay time of τ is the sum of the targets, with an isorange with a maximum delay time of T_p:

$$s_\tau(\tau) = \int_{i\in\text{Footprint}} A_{0i}\delta(\tau_i) \otimes s_t(\tau)\, di \tag{7.2}$$

with A_{0i} denoting the amplitude of the ith target, $\delta(\tau)$ the point-spread function (PSF) of the SAR system, and $\otimes$ the convolution operation. For simplicity of discussion, let us consider a single target without loss of generality. The received signal takes the form

$$s_r(\tau) = A_0 \delta(\tau) \otimes s_t(\tau)$$

$$= \int_0^{T_p} A_0 \delta(t) s_t(\tau - t) \, dt \qquad (7.3)$$

$$= A_0 \exp\left\{ j2\pi \left[f_c(t-\tau) + \frac{1}{2} a_r(t-\tau)^2 \right] \right\}$$

Before the analog-to-digital conversion (ADC) sampling, the received signal is down converted to intermediate frequency (IF):

$$s_{if}(t) = s_t(\tau) s_r(t)^*$$

$$= A_0 \exp\left\{ j2\pi \left[-f_c(t-\tau) - \frac{1}{2} a_r(t^2 - 2t\tau + \tau^2) + f_c t + \frac{1}{2} a_r t^2 \right] \right\} \qquad (7.4)$$

$$= A_0 \exp\left\{ j2\pi \left[f_c \tau + a_r t\tau - \frac{1}{2} a_r \tau^2 \right] \right\}$$

where * denotes the complex conjugate.

Referring to the observation geometry of Figure 7.1 and considering the time delay by the range that is both range and azimuthal position dependent, we extend Equation 7.4 into a two-dimensional expression while taking antenna patterns g_r, g_a into account:

$$s_{if}(\tau, \eta_{m'}) = \int_{m'-L_s(R_0)/2}^{m'+L_s(R_0)/2} \delta(\tau, \eta) g_r(\tau) g_a(\eta)$$

$$\times \exp\left\{ j2\pi \left[f_c \frac{2R(\eta)}{c} + a_r t \frac{2R(\eta)}{c} - \frac{2a_r}{c^2} R(\eta)^2 \right] \right\} d\eta \qquad (7.5)$$

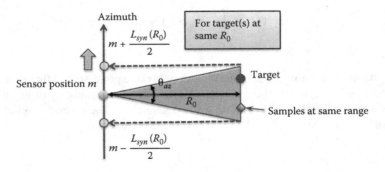

FIGURE 7.1   Illustration of a stationary FMCW observation geometry.

for a distributed target and

$$s_{if}(\tau, \eta_m) = g_r(\tau) g_a(\eta_m) \exp\left\{ j2\pi \left[f_c \frac{2R(\eta_m)}{c} + a_r t \frac{2R(\eta_m)}{c} - \frac{2a_r}{c^2} R(\eta_m)^2 \right] \right\}$$

(7.6)

for a single target, where the delay time and time-dependent range are, respectively,

$$\tau = \frac{2R(\eta, \eta_m)}{c}$$

(7.7)

$$R(\eta, \eta_m) = \sqrt{R_0^2 + u^2(\eta + \eta_m)^2}$$

(7.8)

where $R_0(t)$ corresponds to the shortest distance along the slant range, and m is the azimuth sampling index with $m \in \mathbb{Z}$. The slant range resolution is determined by the inverse of bandwidth $B_{rg} = \frac{a}{\delta f_t}$, while the sampling spacing Δr is dependent on the sampling frequency [2,4,5]:

$$\Delta r = \frac{c}{2\alpha_{os,rg} B_{rg}}$$

(7.9)

where $\alpha_{os,rg}$ is the oversampling rate, $\alpha_{os,rg} > 1$.

The azimuthal bandwidth within the beamwidth θ_{az} is [2]

$$B_{az} = \frac{4 \tan(\theta_{az}/2)}{\lambda}$$

(7.10)

The corresponding azimuth resolution is

$$\rho_{az} = \frac{1}{B_{az}} = \frac{\lambda}{4 \tan(\theta_{az}/2)}$$

(7.11)

The sampling spacing along the slant range is $\frac{c}{2f_s}$, with the sampling frequency constrained by $f_s \geq B_{rg}$. The sampling spacing or the pixel spacing along the azimuthal direction may be defined as

$$\Delta y = \frac{1}{\alpha_{os,az} B_{az}}$$

(7.12)

with $\alpha_{os,az}$ the azimuthal sampling rate, $\alpha_{os,az} > 1$.

Keep in mind that the maximum detectable range for a linear frequency modulation (LFM) system is

$$R_{max} = \frac{cf_s}{4|a_r|} \tag{7.13}$$

The quadratic phase term on the right-hand side of Equations 7.5 and 7.6 is called the residual video phase (RVP) [2–7] and, because the denominator is the speed of light squared, may be ignored in close-range sensing systems. Using Taylor series expansion about $t = 0$, the slant range may be approximated as

$$R(\eta, \eta_m) = \sqrt{R_0^2 + u^2(\eta + \eta_m)^2} \cong \sqrt{R_0^2 + (u\eta_m)^2} + \frac{u^2\eta_m}{\sqrt{R_0^2 + (u\eta_m)^2}}\eta = R_m + \frac{u^2\eta_m}{R_m}\eta \tag{7.14}$$

where

$$R_m = \sqrt{R_0^2 + (u\eta_m)^2} \cong R_0 + \frac{(u\eta_m)^2}{2R_0} \tag{7.15}$$

The range–Doppler frequency is

$$f_d = -\frac{2u_r}{\lambda} = -\frac{2}{\lambda}\frac{\partial R(t, t_m)}{\partial t} = -\frac{2}{\lambda}\frac{u_r^2 t_m}{R_m} \tag{7.16}$$

where u_r is radial velocity.

The slant range may be written in terms of the Doppler frequency as

$$R(\eta, \eta_m) = R_m - \frac{\lambda}{2}f_d\eta \tag{7.17}$$

The delay time in Equation 7.7 may be approximated as

$$\tau = \frac{2R(\eta, \eta_m)}{c} = \frac{2R_m}{c} - \frac{\lambda}{c}f_d\eta \tag{7.18}$$

Now by substituting $R(t, t_m)$ of Equation 7.17 into Equation 7.6 and rearranging the expression, we come up with the following IF signal:

$$s_{if}(\eta, \eta_m) = \exp\left\{j2\pi\left[f_c\left(\frac{2R_m}{c} - \frac{\lambda}{c}f_d\eta\right) + a_r t\left(\frac{2R_m}{c} - \frac{\lambda}{c}f_d\eta\right) - \frac{2a_r}{c^2}R_m^2\right.\right.$$
$$\left.\left. + \frac{2a_r}{c^2}R_m f_d\lambda\eta - \frac{a_r f_d^2\lambda^2}{2c^2}\eta^2\right]\right\} \tag{7.19}$$

For a complete stationary SAR sensor, the range–Doppler frequency is zero, as evidenced by

$$f_d = -\frac{2u_r}{\lambda} = -\frac{2}{\lambda}\frac{\partial R(t_m)}{\partial t} = 0 \tag{7.20}$$

7.3 IMAGE FOCUSING OF SFMCW SYSTEM

As detailed in Chapter 6, there exist numerous focusing algorithms. Because of common and suitable usage, only the range–Doppler algorithm (RDA) [1,2,4,5] and chirp scaling algorithm (CSA) [2,8,9] are adopted to perform the image focusing for the SFMCW system.

7.3.1 FOCUSING BY RDA

As shown in Chapter 6, the core operations in RDA are to perform Fourier transform in the range and azimuthal directions, followed by the range cell migration correction (RCMC). The RCMC is a heavy computation for performing interpolation to track the time dependence of range. After the RCMC operation, an azimuthal matched filter is applied to finish the focusing. Unlike fast-moving platforms, the stationary FMCW system poses essentially little problem on squint angle correction, and hence the estimation of the Doppler centroid. In what follows, let's begin with the Fourier transform in the range direction of Equation 7.19:

$$\begin{aligned} S_{IF}(f_\eta, \eta_m) &= \int s_{if}(\eta, \eta_m)\exp[-j2\pi f_\eta \eta]\,d\eta \\ &= G_r(f_t)g_a(t_m - t_0)\delta\left[f_t - a_r\frac{2R(t_m)}{c}\right]\exp\left\{j2\pi f_c\frac{2R_0}{c}\right\} \\ &\times \exp\left\{j2\pi f_c\frac{(m\Delta y)^2}{cR_0}\right\} \end{aligned} \tag{7.21}$$

In obtaining Equation 7.21, we make use of the point-spread function to approximate $R_0(f_t) = R_0$.

To estimate the beat frequency f_t, note the phase term in Equation 7.21:

$$\psi = 2\pi a_r\frac{2R_0}{c}t + 2\pi a_r\frac{(m\Delta y)^2}{cR_0} - 2\pi f_t t \tag{7.22}$$

By means of the principle of the stationary phase (see appendix in Chapter 5), we have

$$\frac{d\psi}{dt} = 2\pi a_r\frac{2R_0}{c} - 2\pi f_t = 0 \tag{7.23}$$

The beat frequency is found to be

$$f_b = a_r \frac{2R_0}{c} = a_r \tau_0 \tag{7.24}$$

where τ_0 is the time delay from the near range.

Now the Fourier transform in the azimuthal direction, after applying the principle of the stationary phase, is as follows:

$$S_{IF}(f_t, f_m) = \int S_{IF}(f_t, t_m) \exp\{-j2\pi f_m t_m\} dt_m$$

$$= G_r(f_t) G_a(f_m - f_{m_0}) \delta \left\{ \left[f_t - \frac{2aR_{rd}(f_m)}{c} \right] \right\} \exp \left\{ j2\pi f_c \frac{2R_0}{c} \right\} \exp \left\{ -j\pi \frac{f_m^2}{\alpha_0} \right\} \tag{7.25}$$

where

$$\alpha_0 = \frac{2f_c \Delta y^2}{cR_0} \tag{7.26}$$

It is recognized from Equation 7.25 that $R_{rd}(f_m)$ is the Fourier transform of $R_0(\eta_m)$:

$$R_{rd}(f_m) = R_0 + \frac{\Delta y^2}{2R_0 \alpha_0^2} f_m^2 \tag{7.27}$$

From the above equation, we may define the residual terms to be removed through RCMC.

$$\Delta R_{RMC} = \frac{\Delta y^2}{2R_0 \alpha_0^2} f_m^2 \tag{7.28}$$

To complete the process of azimuthal compression, a matched filter is needed to remove the undesired phase term associated with the frequency f_m in Equation 7.25. This can be achieved by designing a filter with a frequency response:

$$H_{az}(f_m) = \exp \left\{ j\pi \frac{f_m^2}{\alpha_0} \right\} \tag{7.29}$$

The matched filtering is of the form

$$S_{ac}(f_t, f_m) = \int S_{IF}(f_t, f_m) H_{az}(f_m) \exp\{-j2\pi f_m t_m\} df_m$$

$$= G_r(f_t) G_a(f_m - f_{m_0}) \delta\{[f_t - f_b]\} \delta[t_m - t_{m_0}] \exp\left\{ j2\pi f_c \frac{2R_0}{c} \right\}$$

$$\times \exp\{j2\pi f_{m_0} t_m\} \tag{7.30}$$

It is shown from the Equation 7.30 that the phase depends on R_0, the shortest range known, and a constant frequency f_{m_0}.

Before proceeding to the next sensation, it is useful to compare the conventional continuous-wave CW SAR and SCW SAR as far as focusing is concern. Table 7.1 lists the parameter differences, including the IF signal, instantaneous slant range, range–Doppler domain signal, and RCMC, between CW and SCWSAR systems.

7.3.2 FOCUSING BY CSA

The chirp scaling algorithm (CSA), proposed by [2,8,9], is to relieve the computation burden in RDA in the context of interpolation by introducing a range-dependent phase function. To better illustrate the idea behind the CSA, we perform a simple simulation of range compression with and without chirp scaled. The original signal after multiplying by a scaling function is plotted in Figure 7.2. Note that only the real part of the signal is displayed. Range compression results are compared at the bottom of Figure 7.2. It is clear that the range migration can be well accounted for by point-wise operation. It is expected that the operation complexity may be reduced from $O(n^2)$ to $O(n^2)$, where n is the number of points to be corrected. Nevertheless, the price to pay in CSA is the large memory required for two-dimensional processing.

Recall from Equation 7.30 that the range-compressed signal is expressed in an azimuthal frequency with the range function

$$R_{rd}(f_m) \leftarrow R_{rd}(R_0, f_m) \approx R_0 + \frac{\lambda^2 R_0 f_m^2}{8} \tag{7.31}$$

Theoretically, the total range migration to be corrected with reference to $f_{m_{ref}}$ is approximately

$$\text{RCM}_{\text{total}}(R_0, f_m) := R_{rd}(R_0, f_m) - R_{rd}(R_0, f_{m_{ref}}) \approx \frac{\lambda^2 R_0}{8}\left(f_m^2 - f_{m_{ref}}^2 \right) \tag{7.32}$$

From [2], of the total range migration, the global or bulk part takes the following expression:

$$\text{RCM}_{\text{bulk}}(f_m) := R_{\text{total}}(R_{\text{ref}}, f_m) \approx \frac{\lambda^2 R_{\text{ref}}}{8}\left(f_m^2 - f_{m_{ref}}^2 \right) \tag{7.33}$$

TABLE 7.1
Parameters of CW and SCW SAR Systems

	CW SAR	SCW SAR
IF signal	$S_{if}(\tau,\eta) = g_a(\eta)\exp\left\{j2\pi\left[f_c\dfrac{2R(\eta)}{c} + a_r\tau\dfrac{2R(\eta)}{c}\right]\right\}$	$S_{if}(\tau,t_m) = g_a(t_m)\exp\left\{j2\pi\left[f_c\dfrac{2R(t_m)}{c} + a_r\tau\dfrac{2R(t_m)}{c}\right]\right\}$
Instantaneous slant range	$R(\eta) = \sqrt{R_0^2 + u^2(\eta-\eta_0)^2} \approx R_0 + \dfrac{u^2(\eta-\eta_0)^2}{2R_0}$	$R(\eta) = \sqrt{R_0^2 + u^2(\eta-\eta_0)^2} \approx R_0 + \dfrac{u^2(\eta-\eta_0)^2}{2R_0}$
Range–Doppler domain signal	$S_{rd}(f_\tau,f_\eta) = \delta\left(f_\tau - \dfrac{2a_r R_{rd}(\eta)}{c}\right)$ $\times \exp\left\{j2\pi f_c\dfrac{2R_0}{c}\right\}\exp\left\{-\pi\dfrac{f_\eta^2}{a_r}\right\}; a_r = \dfrac{2u^2}{\lambda R_0}$	$S_{rd}(f_\tau,f_m) = \delta\left(f_\tau - \dfrac{2a_r R_{ref}(t_m)}{c}\right)$ $\times \exp\left\{j2\pi f_c\dfrac{2R_0}{c}\right\}\exp\left\{-j\pi\dfrac{f_m^2}{\alpha_0}\right\}; \alpha_0 = \dfrac{2}{\lambda R_0}$
RCMC	$R_{rd}(f_\eta) = R_0 + \dfrac{\lambda^2 R_0}{8u}f_\eta^2 = R_0 + \Delta R(f_\eta)$	$R_{ref}(f_m) = R_0 + \dfrac{\lambda^2 R_0}{8}f_m^2 = R_0 + \Delta R(f_m)$

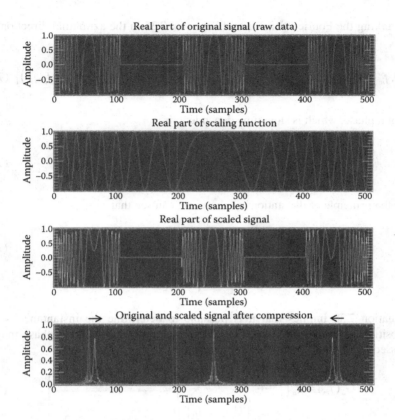

FIGURE 7.2 (See color insert.) Basics of a chirp scaling operation.

while the differential part is

$$RCM_{diff}(R_0, f_m) := R_{total}(R_0, f_m) - R_{total}(R_{ref}, f_m)$$

$$\approx \frac{\lambda^2 R_0}{8}\left(f_m^2 - f_{m_{ref}}^2\right) - \frac{\lambda^2 R_{ref}}{8}\left(f_m^2 - f_{m_{ref}}^2\right) \quad (7.34)$$

It is necessary to know the received signal in the two-dimensional frequency domain in order to obtain the parameters required by the CSA. To do so, we perform the Fourier transform of the range-compressed signal in Equation 7.21:

$$s_{if}\left(\tilde{f}_t, t_m\right) = \int s_{if}(f_t, t_m)\exp\left[-j2\pi\tilde{f}_t f_t\right]df_t$$

$$= G_r\left(\tilde{f}_t\right)g_a(t_m - t_{m_0})\exp\left\{-j2\pi\left[\left(f_c + \tilde{f}_t\right)\frac{2R(t_m)}{c} + a_r\frac{2R(t_m)}{c}\tilde{f}_t\right]\right\} \quad (7.35)$$

where $\tilde{f}_t \overset{FT}{\Longleftrightarrow} f_t$ is a Fourier transform pair.

By taking the Fourier transform of Equation 7.35 in the azimuthal direction, we have

$$S_{\mathrm{IF}}\left(\tilde{f}_t, f_m\right) = \int s_{if}\left(\tilde{f}_t, \eta_m\right)\exp\{-j2\pi f_m \eta_m\}d\eta_m = G_r\left(\tilde{f}_t\right)G_a(f_m)\exp\left\{j\zeta\left(\tilde{f}_t, f_m\right)\right\} \quad (7.36)$$

The total phase, which is slow time dependent, is

$$\zeta(\eta_m) = \frac{-4\pi\left(f_c + \tilde{f}_t\right)R(\eta_m)}{c} - \frac{-4\pi a_r R(\eta_m)}{c} - 2\pi f_m \eta_m \quad (7.37)$$

By the principle of the stationary phase, we can see that

$$\eta_m = \frac{-cR_0 f_m}{2\left[f_c + (1+a_r)\tilde{f}_t\right]\sqrt{1 - \dfrac{c^2 f_m^2}{4\left[f_c + (1+a_r)\tilde{f}_t\right]^2}}} \quad (7.38)$$

Equation 7.38 implies that it is possible to determine the instantaneous sensor position. As such, the azimuthal antenna pattern and the total phase may be expressed in terms of Equation 7.38, giving

$$G_a(f_m) = g_a\left(\frac{-cR_0 f_m}{2\left[f_c + (1+a_r)\tilde{f}_t\right]\sqrt{1 - \dfrac{c^2 f_m^2}{4\left[f_c + (1+a_r)\tilde{f}_t\right]^2}}}\right) \quad (7.39)$$

$$\zeta\left(\tilde{f}_t, f_m\right) = -\frac{4\pi R_0\left[f_c + (1+a_r)\tilde{f}_t\right]}{c}\sqrt{1 - \frac{c^2 f_m^2}{4\left[f_c + (1+a_r)\tilde{f}_t\right]}}$$

$$= \frac{-4\pi R_0 f_c}{c}\sqrt{D^2(f_m) + \frac{2(1+a_r)\tilde{f}_t}{f_c} + \frac{(1+a_r)^2 \tilde{f}_t^2}{f_c^2}} \quad (7.40)$$

In the Equation 7.40, the displacement factor D [2] is

$$D(f_m) = \sqrt{1 - \frac{c^2 f_m^2}{4 f_c^2}} \quad (7.41)$$

Finally, the signal in the frequency domain can be obtained by

$$s_{if}\left(\tilde{f}_t, f_m\right) = \int S_{\mathrm{IF}}\left(\tilde{f}_t, t_m\right)\exp\{j2\pi f_m t_m\}dt_m = G_r\left\{\tilde{f}_t\right\}G_a(f_m)\exp\left\{j\vartheta\left(\tilde{f}_t, f_m\right)\right\} \quad (7.42)$$

In the Equation 7.42, the phase term is

$$
\begin{aligned}
\vartheta\left(\tilde{f}_t, f_m\right) &= \frac{-4\pi R_0 f_c}{c}\sqrt{D^2(f_m)+\frac{2(1+a_r)\tilde{f}_t}{f_c}+\frac{(1+a_r)^2\tilde{f}_t^2}{f_c^2}}+2\pi\tilde{f}_t f_t \\
&= \frac{-4\pi R_0 f_c}{c}\left[D(f_m)+\frac{(1+a_r)\tilde{f}_t}{f_c D(f_m)}-\frac{(1+a_r)^2\tilde{f}_t^2}{2f_c^2 D^3(f_m)}\frac{c^2 f_m^2}{4f_c^2}\right]+2\pi\tilde{f}_t f_t
\end{aligned}
\tag{7.43}
$$

Again, upon using the principle of the stationary phase and after some mathematical manipulations, we reach the following relation:

$$
\tilde{f}_t = \frac{-1}{\beta_m(1+a_r)^2}\left[f_t-\frac{2(1+a_r)R_0}{cD(f_m)}\right]
\tag{7.44}
$$

where

$$
\beta_m = \frac{cR_0 f_m^2}{2f_c^3 D^3(f_m)}
$$

Substituting Equation 7.44 into Equation 7.43, we can readily derive the first-order scaling function as [2]

$$
s_{sc}(f_t, f_m) = \exp\left\{j\pi a_m\left[\frac{D(f_{m_{\text{ref}}})}{D(f_m)}-1\right]f_t^2\right\}
\tag{7.45}
$$

Note that the chirp rate a_r is scaled to a_m through the relation

$$
a_m = \frac{1}{\beta_m(1+a_r)}
\tag{7.46}
$$

Now multiplying Equation 7.21 by the chirp scaling function given in Equation 7.45, we obtain the range migration corrected signal, through chirp rate scaling, as

$$
\begin{aligned}
S_2\left(\tilde{f}_t, f_m\right) &= G_r\left(\tilde{f}_t\right)G_a(f_m-f_{m0})\exp\left\{-j\frac{4\pi R_0 f_c D(f_m)}{c}\right\}\exp\left\{-j\frac{\pi D(f_m)}{a_m D(f_{m_{\text{ref}}})}\tilde{f}_t^2\right\} \\
&\times\exp\left\{-j\frac{4\pi R_0}{cD(f_{m_{\text{ref}}})}\tilde{f}_t\right\}\exp\left\{-j\frac{4\pi}{c}\left[\frac{1}{D(f_m)}-\frac{1}{D(f_{m_{\text{ref}}})}\right]R_{\text{ref}}\tilde{f}_t\right\} \\
&\times\exp\left\{j\frac{4\pi a_m}{c^2}\left[1-\frac{D(f_m)}{D(f_{m_{\text{ref}}})}\right]\left[\frac{R_0}{D(f_m)}-\frac{R_{\text{ref}}}{D(f_m)}\right]^2\right\}
\end{aligned}
\tag{7.47}
$$

TABLE 7.2
Focusing Parameters and Functions between CW and SCW SAR Systems

	CW SAR	SCW SAR
Displacement factor	$D(f_\eta, u) = \sqrt{1 - \dfrac{c^2 f_\eta^2}{4u^2 f_c^2}}$	$D(f_m, u) = \sqrt{1 - \dfrac{c^2 f_m^2}{4 f_c^2}}$
Modified range FM rate	$a_m = \dfrac{a_r}{1 - a_r \left(\dfrac{cR_0 f_\eta^2}{2u^2 f_c^3 D^3(f_\eta, u)} \right)}$	$a_m = \dfrac{a_r}{1 - a_r \left(\dfrac{cR_0 f_m^2}{2 f_c^3 D^3(f_m)} \right)}$
First-order scaling function	$S_{sc1}(f_\tau, f_\eta) = \exp\left\{ j\pi a_m \left[\dfrac{D(f_{\eta_{ref}}, u_{ref})}{D(f_\eta, u_{ref})} - 1 \right] (f_\tau)^2 \right\}$	$S_{sc1}(f_\tau, f_m) = \exp\left\{ j\pi a_m \left[\dfrac{D(f_{m_{ref}})}{D(f_m)} - 1 \right] (f_\tau)^2 \right\}$
Second-order scaling function	$S_{sc2}(t', f_\eta) = \exp\left\{ -j \dfrac{\pi D(f_\eta, u)}{a_m D(f_{\eta_{ref}}, u)} t'^2 \right\}$ $\times \exp\left\{ -j \dfrac{4\pi}{c} \left[\dfrac{1}{D(f_\eta, u_{ref})} - \dfrac{1}{D(f_{\eta_{ref}}, u_{ref})} \right] R_{ref} t' \right\}$	$S_{sc2}(t', f_m) = \exp\left\{ -j \dfrac{\pi D(f_m)}{a_m D(f_{m_{ref}})} t'^2 \right\}$ $\times \exp\left\{ -j \dfrac{4\pi}{c} \left[\dfrac{1}{D(f_m)} - \dfrac{1}{D(f_{m_{ref}})} \right] R_{ref} t' \right\}$
Azimuth matched filter	$H_{az}(R_0, f_\eta) = \exp\left\{ j \dfrac{4\pi R_0 f_c D(f_\eta, u)}{c} \right\}$	$H_{az}(R_0, f_m) = \exp\left\{ j \dfrac{4\pi R_0 f_c D(f_m)}{c} \right\}$

The first phase term is a form of the azimuthal-compressed frequency-modulated signal and is dependent on the range R_0. The second phase term, an azimuth dependent, is the residual term after range compression and thus is undesired. This term is commonly ignored in RDA. The third phase term is useful in interferometric processing. A critical part is the fourth phase term, a global range migration correction term to compensate the range-varying phase. The last phase term is to compensate the residual error in azimuthal compression. In cases of spaceborne sensors, this term may be ignored. Table 7.2 summarizes the key parameters and scaling functions used for image focusing in CW and SCW SAR systems.

7.4 MOTION COMPENSATION

The motion compensation algorithms were presented in Chapter 6 for the case of pulse SAR. Even in a stationary CW system, a sensor's position and motion need to be seriously taken into account in order to achieve a high-quality image focus. The stationary FMCW system, though limited in practical remote sensing, offers a powerful tool to understand the wave–target interaction mechanisms, explore the remote sensing science, and design a new SAR system to fulfill the mentioned tasks. In this section, we consider the motion compensation in RDA and CSA focusing described in the previous sections [2,6–8,10–12]. To facilitate the following simulation, we begin with the signal model for the first and second MoCo. From Equation 7.15, we may express the actual range function as a sum of an ideal range, $R(t_m)$, and a residual range:

$$R(t_m) \rightarrow R(t_m) + \Delta R(t_m) \tag{7.48}$$

where $\Delta R(t_m)$, the term to be compensated, is the fluctuation error between the ideal and actual range, which might be recorded during the data acquisition.

Now, we replace the range function in Equation 7.6 with Equation 7.48 to obtain

$$s_{if}(\tau, \eta_m) = g_r(\tau) g_a(\eta_m) \exp\left\{ j2\pi \left[f_c \frac{2R(\eta_m)}{c} + a_r \tau \frac{2R(\eta_m)}{c} - \frac{2a_r}{c^2} R(\eta_m)^2 \right] \right\}$$

$$\times \exp\left\{ -j2\pi \left[f_c \frac{2\Delta R(\eta_m)}{c} + a_r \tau \frac{2\Delta R(\eta_m)}{c} - \frac{4a_r}{c^2} R(\eta_m) \Delta R(\eta_m) \right. \right.$$

$$\left. \left. + \frac{2a_r}{c} \Delta R(\eta_m)^2 \right] \right\} \tag{7.49}$$

The phase to be compensated is identified as

$$\phi_{MC} = \exp\left\{ -j2\pi \left[f_c \frac{2\Delta R(\eta_m)}{c} + a_r \tau \frac{2\Delta R(\eta_m)}{c} - \frac{4a_r}{c^2} R(\eta_m) \Delta R(\eta_m) \right. \right.$$

$$\left. \left. + \frac{2a}{c^2} \Delta R(\eta_m)^2 \right] \right\} \tag{7.50}$$

The essential part now is to find $\Delta R(\eta_m)$. Practically, we may decompose the displacement, $\vec{d}$, into three components under the line-of-sight, perpendicular, velocity (LPV) coordinate system: the line of sight $\hat{d}_{los}$, the parallel component that is along the azimuthal direction, and a perpendicular component $\hat{d}_\perp$ that is orthogonal to $\hat{d}_{los}$ and $\hat{d}_m$ (see Figure 4.10). It follows that the error term can be expressed in terms of the three LPV components as

$$\Delta R(t_m) = d_{los} - \frac{d_{los}}{2R_0^2}(t_m - t_{m_0} + d_m)^2 - \frac{d_\perp^2}{2R_0} - \frac{d_{los}}{2R_0^2}d_\perp^2 - \frac{(t_m - t_{m_0})d_m}{R_0} - \frac{d_m^2}{2R_0} \quad (7.51)$$

It is seen that the major contribution comes from d_{los}; other terms may be ignored in one way or the other, depending on the sensor trajectory stability. For the ground-based stationary SAR, component d_m is small, and $d_\perp$ is relatively small compared to d_{los}. Following the definition in Chapter 4 and according to a desired SAR observation geometry, such as the one considered here, it might be convenient to define a SAR coordinate in the slant range plane

$$\begin{cases} \hat{v}_m = [0,1,0] \\ \hat{v}_{los} = \text{VecFindArbitraryRotate}\left([0,0,-1], \theta_\ell, \hat{v}_m\right) \\ \hat{v}_\perp = \hat{v}_{los} \times \hat{v}_m \end{cases} \quad (7.52)$$

Now denoting $\vec{d}$ as the displacement vector in the LPV coordinate, the next step is to transform the LPV coordinate to SAR coordinates by projection obtained by

$$\begin{cases} d_{los} = \vec{d} \cdot \hat{v}_{los} \\ d_m = \vec{d} \cdot \hat{v}_m \\ d_\perp = \vec{d} \cdot \hat{v}_\perp \end{cases} \quad (7.53)$$

where d components are defined in the slant range plane, as described in Section 4.5.3.

The first motion compensation term is

$$\phi_{MC1} = \exp\left\{-j2\pi\left[f_c\frac{2R_{ref}(t_m)}{c} + a_r t\frac{2R_{ref}(t_m)}{c}\right]\right\} \quad (7.54)$$

This involves the azimuthal-dependent slant range $R_{ref}(t_m)$ and is a one-dimensional operation. Notice that the second term of the right-hand side in the Equation 7.54 is usually ignored in pulse SAR systems because t is small.

The second motion compensation is the phase of the form

$$\phi_{MC2} = \exp\left\{-j2\pi\left[2f_c\frac{\Delta R(R_0,t_m) - R_{ref}(t_m)}{c}\right]\right\} \quad (7.55)$$

The slant range to be corrected is $\Delta R(R_0, t_m)$. Unlike the first MoCo, the second MoCo is a two-dimensional operation because it is performed after the range compression.

In what follows, numerical simulation is performed to illustrate the motion compensation for a stationary FMCW system. Table 7.3 summarizes the sensor parameters and the associated imaging parameters. The motion bias was to 5 cm. With a carrier frequency of 90 GHz, it is expected that the vibration-induced phase error would be large, at least not negligible. Figure 7.3 shows the 10 point targets' arrangement in the ground and converted slant range planes, where the solid circles indicate the point target being imaged with the (x, y) and (y, r) values included. The above

TABLE 7.3
Simulation Parameters for SCW SAR System

Parameter	Value
Carrier frequency	90 GHz
Transmitting bandwidth	13 GHz
Duration time	0.05 µs
Receiving sampling rate	15 GHz
Sensor position interval	0.004 m
Sensor height	1.5 m
Antenna beamwidth at azimuth	15°
Antenna beamwidth at elevation	35°
Look angle	50°
Motion bias standard deviation	0.05 m
Slant range resolution	1.6689 cm
Azimuth resolution	0.4723 cm
Slant range pixel spacing	1.2491 cm
Azimuth pixel spacing	0.4000 cm

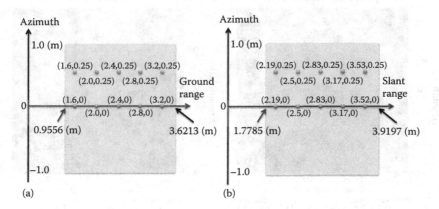

FIGURE 7.3 Target location's arrangement in (a) ground range and (b) slant range, with a total of 10 point targets. The actual location is indicated in units of meters.

setup and arrangement is actually based on a small microwave aniconic chamber devised from SAR imaging measurements.

For simplicity but without loss of generality, we only consider the azimuthal antenna pattern, which is of more interest. The amplitude and phase of the echo signal are displayed in Figure 7.4. The amplitude variation is caused by the range changes, the antenna pattern along the azimuthal direction, and the speckle effect.

In RDA, the range compression was performed using a Kaiser window ($\beta = 2.1$) in fast Fourier transform (FFT). In applying Equation 7.28 for RCMC, a simple sinc interpolator was used with eight samples in kernel. The results are displayed in Figure 7.5, where we can see 10 curves, each corresponding to 10 individual targets; in each

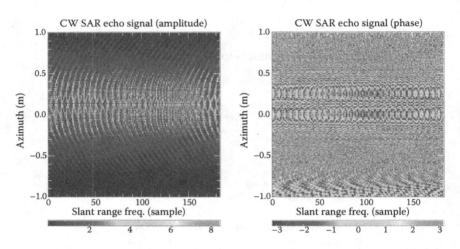

FIGURE 7.4 **(See color insert.)** Echo signal of the imaging scenario given in Figure 7.3 with sensor parameters specified in Table 7.3.

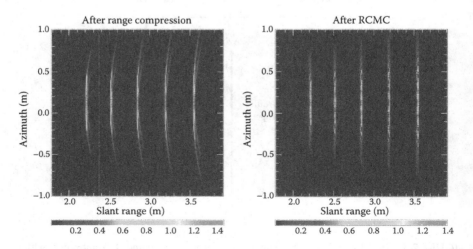

FIGURE 7.5 **(See color insert.)** Simulated data after range compression and after RCMC.

set, the 2 crossed curves represent 2 targets with the same slant range, but at different azimuth positions. The RCMC result is also displayed in Figure 7.5. The correction of the curvature of the range-compressed signal is clearly shown. The coherent sum of the two targets at the same slant range but at different azimuth positions is also observed. After azimuthal compression, these two targets will be differentiated because of the Doppler frequency embedded in the azimuthal frequency domain.

To investigate the sensor motion effect and its compensation, random motion in the ground range and in altitude is simulated by the procedures presented in Chapters 4 and 6. The simulated motion bias is shown in Figure 7.6.

Compared to the original range-compressed image, after motion compensation, it is seen that the response becomes less noisy—the energy is more confined, as evident from Figure 7.7. It is more prominent around the scene center because in the first MoCo, $R_{ref}(t_m)$ is referred to the scene center. Hence, it is expected that the compensation is less effective away from the scene center. After the second MoCo, the degraded performance can be improved as we further consider the phase associated with $\Delta R(R_0, \eta_m) - R_{ref}(\eta_m)$, as given in Equation 7.54. For better visual inspection of such improvements, enlarged portions of the compressed image of the two targets at the far range are selected, as shown in Figure 7.8. It is shown that unlike by only performing the first MoCo, by carrying out the second MoCo, the compensation performance is no longer dependent on the slant range away from the scene center. To further explore the differences between the first and second phase compensation for sensor motion, Figure 7.9 illustrates the focusing effect after the first and second MoCo processes. It is now more evident to see the impact of the first and second MoCo, respectively. Equal performance across the whole scene is obtained after the first and second MoCo.

It is interesting to compare the RDA focusing under the random sensor motion with and without MoCo, as shown in Figure 7.10. Clearly, without motion compensation, the energy is spread along the azimuth direction and very poor azimuthal focusing results. It is actually not acceptable. After first and second MoCo, a well-focused

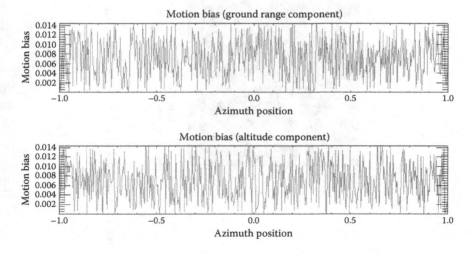

FIGURE 7.6 Simulated random motion in the ground range and in altitude, in meters.

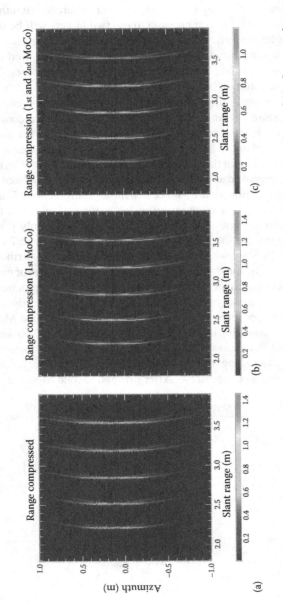

FIGURE 7.7 (See color insert.) Effect of motion compensation, (a) no compensation, (b) after first compensation, and (c) after second compensation.

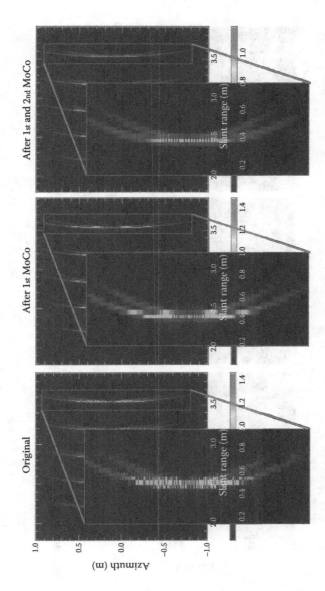

FIGURE 7.8 **(See color insert.)** Enlarged portions of the range-compressed images selected from the two targets at the far range.

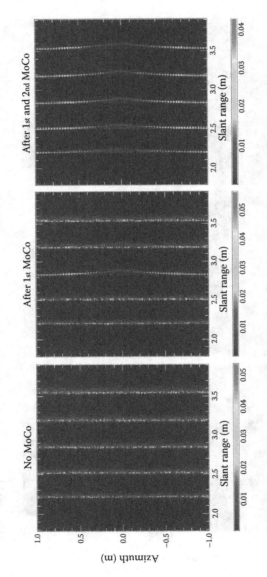

FIGURE 7.9 (See color insert.) Illustration of first and second motion compensation to demonstrate the focusing effect.

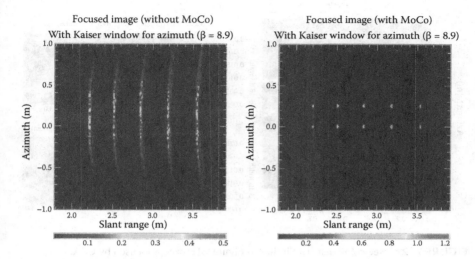

FIGURE 7.10 (See color insert.) Focused image by RDA with and without motion compensation.

image can be obtained. In fact, the focused quality is comparable to that of the motionless focused image, as shown in Figure 6.6. Hence, it can be said that in either data-based or signal-based algorithms, compensation for sensor-induced motion is necessary even for the stationary FMCW system. Before closing, it is worth emphasizing that the simulation procedure presented here is useful to trace possible sources of phase error, such as sensor-induced motion effects.

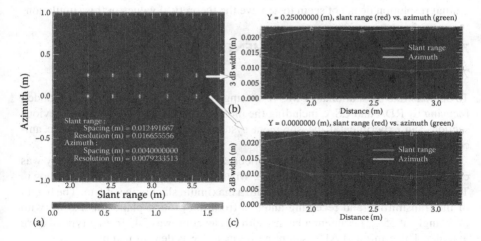

FIGURE 7.11 (See color insert.) Geometric quality of the RDA focused image after MoCo. The upper right plot corresponds to the upper row of targets, and the lower right plot with the lower row of targets. (a) RDA focused image, (b) enlarged upper row target of (a), (c) enlarged upper row target of (a).

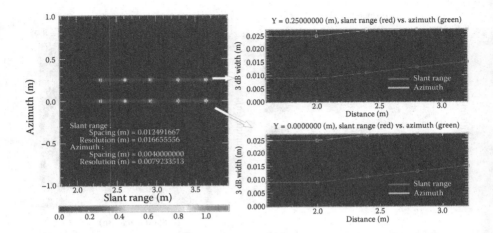

FIGURE 7.12 (**See color insert.**) Similar to Figure 7.11, except focused by CSA.

In applying the azimuth matched filter given by Equation 7.29, a Kaiser window with β = 8.9 was adopted when performing the FFT. Figure 7.11 displays the final RDA focused image after MoCo. The focusing quality is evaluated by computing the 3 dB width for each target. The average slant range resolution was 1.2 cm compared to a theoretical value of 1.67 cm; the average azimuth resolution was 2.2 cm, which is below the ideal value of 0.63 cm. Apparently, a refined procedure must be devised to improve the focusing quality. Nevertheless, the geometric accuracy is quite satisfactory even under a strong sensor motion error. For the same set of simulation parameters, focusing by CSA is also performed. Results are displayed in Figure 7.12. Image quality by CSA is very similar to that of RDA. Similarly, the resulting azimuthal resolution does not seem to achieve the theoretical values in this simulation.

7.5 EXPERIMENTAL MEASUREMENTS

7.5.1 MEASUREMENT SETUP

To demonstrate the stationary FMCW sensing system and validate the signal model, focusing by RDA and CSA, including the motion compensation as given in previous sections, an experimental measurement system was realized in a microwave aniconic chamber.

Based on the size of the available aniconic chamber, the imaging geometry was configured as illustrated in Figure 7.13, where the relevant parameters are shown. The total azimuth length was 2.37 m, with a maximum slant range of 6 m. The height of the transmitting and receiving antenna from the ground plane was 2.5 m, with look angle at 55°. The antenna beamwidth in elevation was 24° from a typical horn antenna. The minimum ADC sampling frequency was determined by

$$f_{s,\min} = \frac{4R_{\max}|a_r|}{c}$$

(7.56)

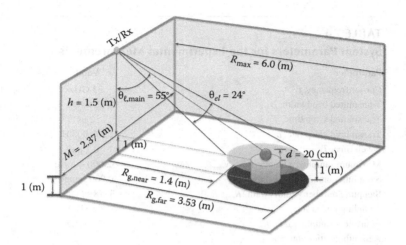

FIGURE 7.13 Experimental measurement setup.

The movable length for the sensor was constrained by the following relation:

$$y_{max} = d + 2R_{g,far} \tan\left(\frac{\theta_{az}}{2}\right) \qquad (7.57)$$

where d is the target diameter.

Table 7.4 lists the system parameters for the experimental measurements. From Equations 7.13 and 7.56, the minimum ADC sampling frequency was 200 Hz. To determine the required azimuthal sampling frequency, it was necessary to take the slant range bandwidth into account. Hence, we can set $f_s = \max[f_{s,min}, B_{rg}]$.

Looking at the raw data (amplitude and phase) in Figure 7.14 (left), a strong but undesired signal appeared somewhere between 0.05 and 0.06 μs delay time. Through examining the measurement setup, the unwanted radiation source was identified as coming into the antenna from the side lobe. The removal of this source was straightforward. The corrected signal is displayed in Figure 7.14 (right). Even at close range in this case, the range migration effect was still visible. Compared to the background floor, the radar cross section (RCS) of the test target (metal sphere) makes the contrast quite large. Hence, even though there exists noise contamination surrounding the target, focusing quality should not be affected.

7.5.2 IMAGE FOCUSING

Theoretically, measurements in a well-controlled chamber and parameters associated with observation geometry are all close to ideal values. Nevertheless, the experiment offers a useful approach to developing and designing a SAR imaging system. From the raw data acquired, the Doppler centroids estimated by multi-look cross correlation (MLCC) and multi-look beat frequency (MLBF) [2] were −9.19 and −8.32 Hz, respectively. The azimuth ambiguity number was zero. This

TABLE 7.4
System Parameters for the Experimental Measurements

Parameter	Value
Carrier frequency, f_c	35 GHz
Transmitted bandwidth, B_{rg}	5 GHz
Transmitted swap time, T_r	2 μs
Transmitted chirp rate, a_r	2,500,000 Hz
Dwell time, T_d	0.1 μs
Maximum sampling number in T_d, N_{max}	32,000 sample
Sampling number in T_d	30,000 samples
Sampling number within swath $[R_{near}, R_{far}]$	3578 samples
Sampling rate at R_{max}	64 MHz
Equivalent sampling rate, $f_{s,eq}$	300 GHz
Azimuth position interval	1.0 cm
Azimuth beamwidth	24°
Elevation beamwidth	24°
Look angle at boresight	55°
Azimuth position range	−1.35–1.35 m
Sensor height	1.5 m
Target location region at ground range	1.40–3.53 m
Target extension at slant range	2.05–3.83 m
Target extension at azimuth	0.30–2.70 m
Maximum target diameter of target	80 cm
Azimuth bandwidth	99.26 Hz
Spatial resolution (slant range, azimuth)	3.00, 1.01 cm
Pixel spacing (slant range, azimuth)	0.05, 1.00 cm

corresponds to a backward squint angle of 2.26°. The azimuth instantaneous frequency is shown in Figure 7.15. The estimated Doppler center frequency of −9.19 Hz, f_{abs}, was centered at the azimuth frequency, whose bandwidth is $1/\delta_{az}$. Because f_{abs} is nonzero, an offset of 22.02 pixels to the left can be estimated accordingly.

To more closely exam the effect of this offset, let's look at the data in the range–azimuth frequency domain, as shown in Figure 7.16a. A fast frequency change occurs around the center frequency, with lower-frequency components around the high-frequency component. The curves crossing was due to a relatively small synthetic aperture size compared to the azimuthal antenna beamwidth. The correct this effect, the necessary range cell migration correction, according to Equation 7.28, is illustrated in Figure 7.16b. The color represents the magnitude in pixels that needed to be corrected in Figure 7.16a, with the negative sign meaning a correction to the left.

As explained above, and upon applying the RCMC in Figure 7.16a and b, the result displayed in Figure 7.17 was obtained. Compared to Figure 7.16a, it is revealed that range cell migration is accurately corrected. As mentioned earlier, noisy signal still remains due to the antenna side lobe, but it should not affect the final focusing.

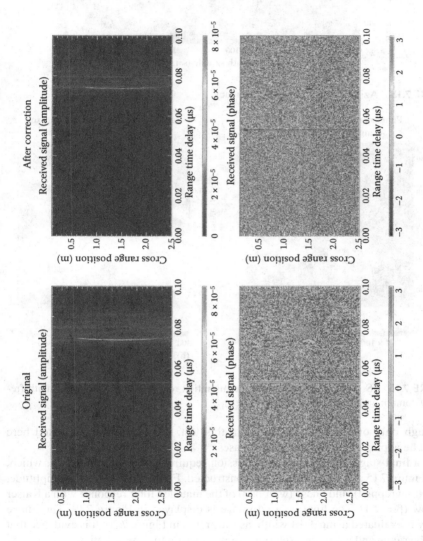

FIGURE 7.14 (See color insert.) Original raw data (amplitude and phase) and corrected data.

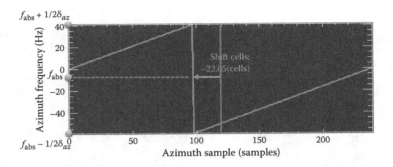

FIGURE 7.15 Azimuth instantaneous frequency.

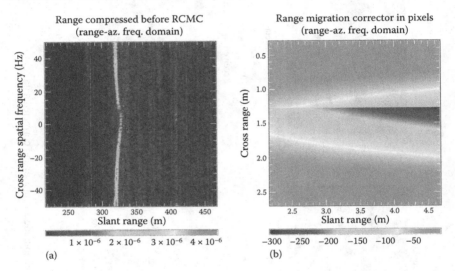

FIGURE 7.16 **(See color insert.)** (a) Range migration revealed in the range–azimuth frequency domain and (b) required RCMC pattern.

Although only one target was placed in this test, the processing presented here should be applicable to multiple-target cases.

As a final stage, the azimuth compression requires a reference function, which, from Figure 7.15 and Equation 7.29, is constructed. Figure 7.18 shows the amplitude, phase (unwrapped), and phase (wrapped) of the matched filter response with a Kaiser window ($\beta = 2.1$). The final focused image is displayed in Figure 7.19. The image quality is evaluated using 3 dB width, as illustrated in Figure 7.20. It is evidence that both the range and azimuth resolution are well within the specifications.

As a final example, imaging a vehicle (Figure 7.21) whose computer-aided drafting (CAD) model was placed on the rotation pedal (Figure 7.13) was performed. The RCS computation follows the procedures described in this chapter. Simulated images for aspect angles from 0° to 360° were also obtained, as shown in Figure 7.22. The real measured image with an aspect angle of 45° was acquired, with the excellent

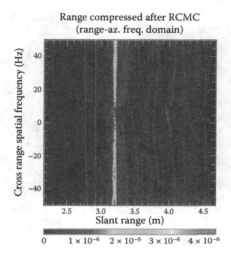

FIGURE 7.17 **(See color insert.)** Data in range–azimuth frequency domain after RCMC.

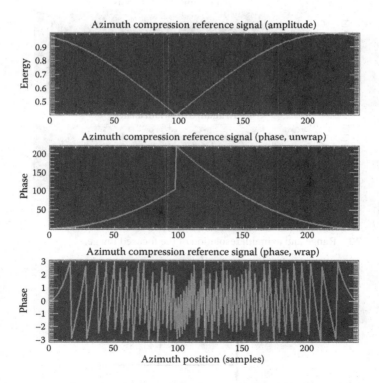

FIGURE 7.18 Response of the azimuth matched filter: amplitude, phase (unwrapped), and phase (wrapped) of the matched filter response with the Kaiser window ($\beta = 2.1$).

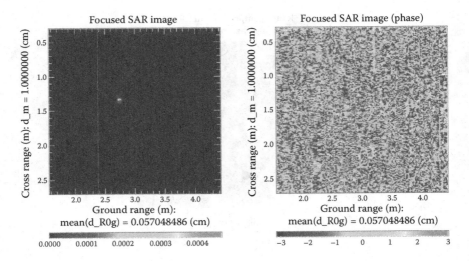

FIGURE 7.19   (See color insert.) Focused image showing both amplitude and phase.

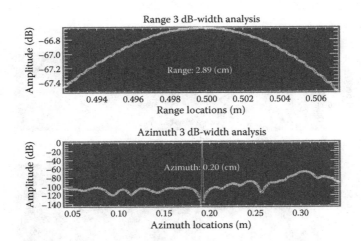

FIGURE 7.20   Range and azimuth resolution of the focused image.

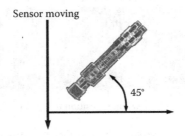

FIGURE 7.21   Target geometry of a CAD model.

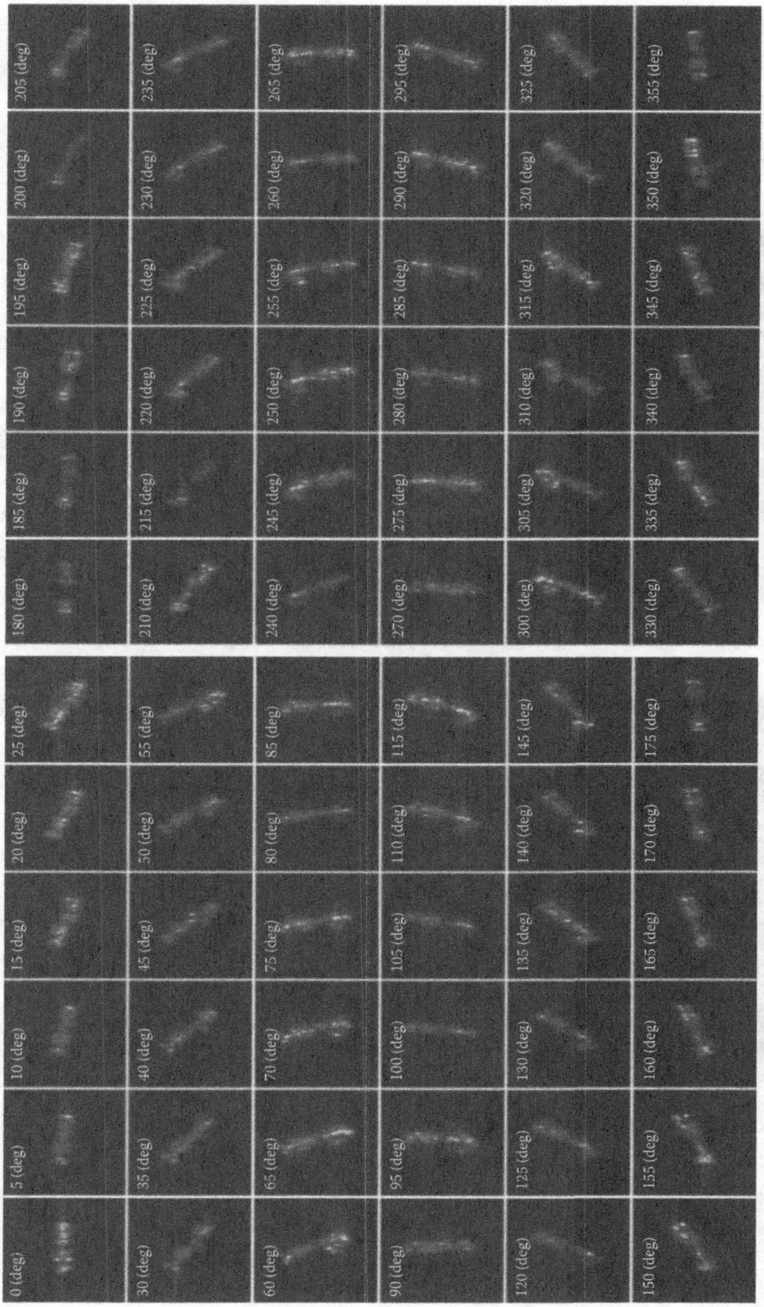

FIGURE 7.22 Simulated images of target model (Figure 7.21) with aspect angles of 0°–360°.

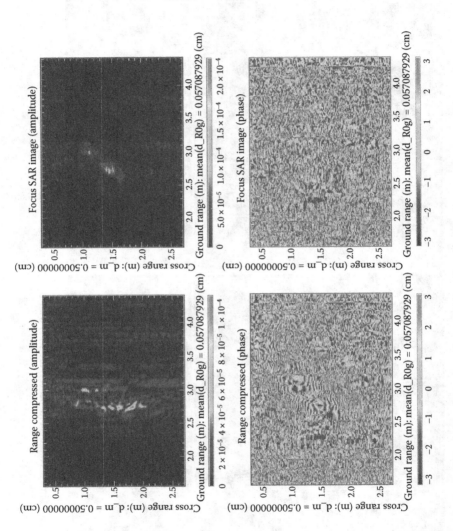

FIGURE 7.23 (See color insert.) Range-compressed image and focused image of the target.

image quality of the focus shown in Figure 7.23, where the range-compressed and focused images, both amplitude and phase, are displayed. The scattering centers at this target geometry are well preserved. Notice that in RCS computation, multiple scattering was not included. Hence, in quantitative comparison between simulated and measured images, there exist differences in where multiple scattering is important. The fidelity can be easily achieved and improved in more accurate RCS computation. Nevertheless, the point here that the simulation and experiment setup given in this section offers a useful and effective approach to studying the SAR system with respect to signal model design, focusing algorithm, and motion compensation.

REFERENCES

1. Carrara, W. G., Majewski, R. M., and Goodman, R. S., *Spotlight Synthetic Aperture Radar: Signal Processing Algorithms*, Artech House, Norwood, MA, 1995.
2. Cumming, I. G., and Wong, F. H., *Digital Processing of Synthetic Aperture Radar Data: Algorithms and Implementation*, Artech House, Norwood, MA, 2005.
3. Kirk, J. C., *Motion Compensation for Synthetic Aperture Radar*, Technology Service Corporation, Los Angeles, CA, 1999.
4. Meta, A., Hoogeboom, P., and Ligthart, L., Signal processing for FMCW SAR, *IEEE Transactions on Geoscience and Remote Sensing*, 45(11): 3519–3532, 2007.
5. Curlander, J. C., and McDonough, R. N., *Synthetic Aperture Radar: Systems and Signal Processing*, Wiley-Interscience, New York, 1991.
6. Franceschitti, G., and Lanari, R., *Synthetic Aperture Radar Processing*, CRC Press, New York, 1999.
7. Kirk, J. C., Motion compensation for synthetic aperture radar, *IEEE Transactions on Aerospace and Electronic Systems*, 3: 338–348, 1975.
8. Moreira, J. R., A new method of aircraft motion error extraction from radar raw data for real time motion, *IEEE Transactions on Geoscience and Remote Sensing*, 28(4): 620–626, 1990.
9. Moreira, A., and Huang, Y., Airborne SAR processing of highly squinted data using a chirp scaling approach with integrated motion compensation, *IEEE Transactions on Geoscience and Remote Sensing*, 32(5): 1029–1040, 1994.
10. Moreira, A., Mittermayer, J., and Scheiber, R., Extended chirp scaling algorithm for air-and spaceborne SAR data processing in Stripmap and ScanSAR imaging modes, *IEEE Transactions on Geoscience and Remote Sensing*, 34(5): 1123–1136, 1996.
11. Raney, R. K., Runge, H., Bamler, R., Cumming, I. G., and Wong, F. H., Precision SAR processing using chirp scaling, *IEEE Transactions on Geoscience and Remote Sensing*, 32: 786–799, 1994.
12. Zaugg, E. C., and Long, D. G., Theory and application of motion compensation for LFM-CW SAR, *IEEE Transactions on Geoscience and Remote Sensing*, 46: 2990–2998, 2008.

8 System Simulations and Applications

8.1 INTRODUCTION

This chapter aims to systematically integrate the relevant topics presented in the previous chapters to perform system simulation, a major theme of this book, so that readers may have a better and deeper understanding of synthetic aperture radar (SAR) principles. As emphasis, SAR is a complex device that involves radar wave interactions with the target being imaged for signal processing and image formation and focusing. Two major figures of merit concerning the image quality are the geometric and radiometric accuracy [1]. Hence, it is highly desirable to apply a full-blown SAR image simulation scheme with high verisimilitude, including the sensor and target geolocation relative to the earth, movement of the SAR sensor, SAR system parameters, radiometric and geometric characteristics of the target, and environment clutter. The simulation scheme should at least include the computation of the target's radar cross section (RCS) or scattering coefficient, trajectory parameter estimation, SAR echo signal generation, and image focusing. Such a simulation scheme is also well suited for spaceborne SAR mission planning. As an application example, identification of a deterministic target over a random background will be demonstrated. The target contrast enhancements by means of nonquadratic optimization and feature extraction are also given to show the complete simulation and application.

8.2 SIMULATION DEMONSTRATION

The simulation working flows that combining target location, RCS computation, speckle and clutter, echo signal generation, image focusing, and motion compensation algorithms were illustrated and validated by evaluating the image quality, including geometric and radiometric accuracy using simple point targets first, followed by simulating a complex target of four types of commercial aircraft: A321, B757-200, B747-400, and MD80. The satellite SAR systems for simulation include Radarsat-2, TerraSAR-X, and ALOS PALSAR, but other systems can be easily realized as long as their system parameters are available.

8.2.1 Point Target Response

To evaluate the simulation performance, the following common figures of merit are used [2–4]:

1. 3 dB width of point-spread function (PSF)
2. Peak side lobe ratio (PSLR) in the range and azimuth directions
3. Integrated side lobe ratio (ISLR)

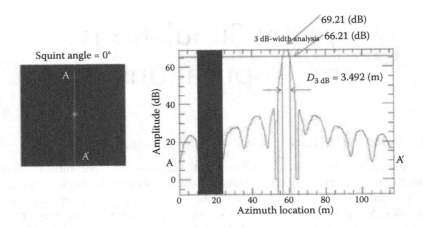

FIGURE 8.1   Simulated TerraSAR image showing the PSF in the azimuth direction.

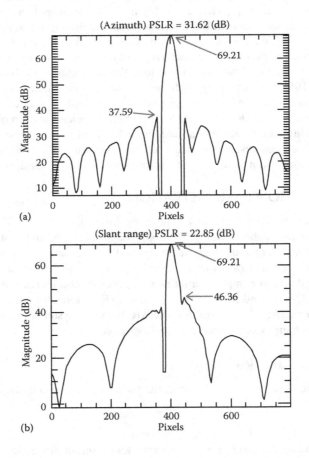

FIGURE 8.2   PSLR of simulated TerraSAR image in (a) the azimuth direction and (b) slant range.

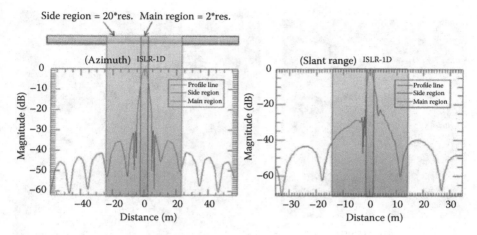

FIGURE 8.3 ISLR of simulated TerraSAR image in the azimuth direction and slant range.

Figure 8.1 is the PSF of a simulated TerraSAR of a point target. Only the azimuth resolution is of more concern here. As shown in the plot, the 3 dB width (spatial resolution) well meets the nominal resolution. The PLSR and ISLR performances are plotted in Figures 8.2 and 8.3, respectively. We can readily check that the PLSR and ISLR along the azimuth and slant range directions are well controlled within the nominal required values. It must be noted that the point target simulation at this moment does not consider the clutter or speckle effect. It is thus expected that at the 3 dB width, PLSR and ISLR will be degraded when the scattering from the background is included. To further demonstrate the simulation procedure, we investigate the target location accuracy.

8.2.2 TARGET LOCATION

As described in Chapter 4 the sensor coordinates and target on the earth's surface must be lined up to uniform coordinates of both time and space. The SAR signal is sensitive to the target aspect angle. The target aspect angle is defined as the angle between the SAR looking direction and the azimuthal direction in the ground plane. Figure 8.4 illustrates a target of an MD80 aircraft setting up with respect to the SAR (e.g., TerraSAR-X) flying path. The scene center was determined from a real image. A total of 11 state vectors from the orbit data file were used for simulation.

A comparison of the locations between real and simulated TerraSAR scenes in both geodetic and earth central rotational (ECR) coordinates is given in Table 8.1. The difference of an order of $10^{-4\circ}$ was obtained. The corresponding ground range difference of 0.06 m and a difference of 16.97 m were observed and are highly acceptable.

Next we consider a point target. Figure 8.5 gives the schematic ALOS PALSAR ascending path state vector starting at 14:15:0.0 Coordinated Universal Time (UTC) on November 30, 2007. A total of 28 state vectors with a duration of 60 s were used in the simulation. Relevant parameters of the sensor are also given in the table. For simplicity, a zero squint angle was considered.

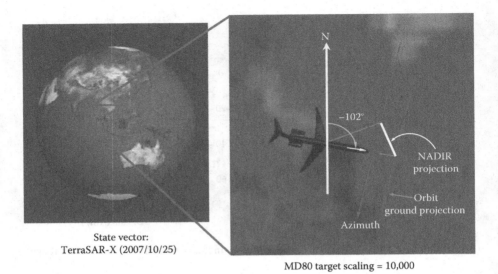

FIGURE 8.4 **(See color insert.)** Target setting on the earth's surface for a TerraSAR image simulation.

TABLE 8.1

Performance of System Simulation for a Point Target Using ALOS PALSAR and TerraSAR-X Systems

Index		ALOS PALSAR		TerraSAR-X	
		Simulation	Nominal	Simulation	Nominal
Azimuth 3 dB beamwidth (m)		4.26	5.10	3.49	4.51
PSLR (dB)	Slant range	−24.85	−20.5	−22.85	−13
	Azimuth	−25.42		−31.16	
ISLR (dB)	Slant range	−30.62	−15.0	−31.62	−17
	Azimuth	−25.62		−43.98	
Scene center	Lon, lat (°)	121.49927° E, 24.764622° N	121.49930° E, 24.764477° N	119.395090° E, 24.894470° N	119.395060° E, 24.894620° N
	ECR (m)	−3027814.2, 4941079.3, 2655407.1	−3027808.4, 4941075.0, 2655421.7	−2817396.77, 5608995.60, 2830631.97	−2817397.45, 5608996.11, 2830630.29
	Lon/lat difference (°)	−3 × 10⁻⁵, 1.45 × 10⁻⁴		−3 × 10⁻⁶, 1.5 × 10⁻⁴	
	ECR difference (ground range, azimuth) (m)	0.98, −16.26		0.06, 16.97	

Parameter	Value
Transmitted freq.	1.27 (GHz)
PRF	2159.8272 (Hz)
Range sampling rate	32 (MHz)
Effective antenna 3 dB beamwidth (elevation, azimuth)	(5.2, 1.3) (degrees)
Chirp bandwidth	28 (MHz)
Pulse duration time	27 (µs)
Antenna mechanical boresight	47.6 (degree)
Earth semimajor axis	6,378,137.0 (m) (GRS80)
Earth semimirror axis	6,356,752.3141 (m) (GRS80)
Time of first state vector	2007/11/30 14:15:0.0 (UTC)
Number of state vector	28
Time interval of state vector	60 (s)
Ascending/descending	Ascending
Simulation squint angle	0 (degree)
Simulation data dimension	(2475, 8814) (pixels)

FIGURE 8.5 System simulation parameters and state vectors from ALOS-PALSAR taken on November 30, 2007, at starting time 14:15:0.0 UTC.

Selected quality indices to evaluate simulated images of TerraSAR-X and PLASAR satellite SAR systems are listed Table 8.1, where quality indices include 3 dB azimuth width (m), peak–side lobe ratio (dB), integrated side lobe ratio (dB), and scene center accuracy in the ECR coordinate. As can be seen from the table, the simulated images following the procedure given in Chapters 4 through 6 are validated and all found to be well within the nominal specifications for different satellite systems.

Figure 8.6 displays several simulated TerraSAR-X image clips of MD80 (Figure 8.6a) and B757 (Figure 8.6b) commercial aircraft with an incident angle of 30° and aspect angles of 0°–54°. A simulated Radarsat-2 image of an A321 aircraft is shown in Figure 8.7. These simulated images will be used to demonstrate the application of target identification, to be discussed in the following sections.

In short, four major steps are needed for system simulation, including taking a satellite orbit into account:

1. Find satellite position, calculated from two-line element data and imaging time duration.
2. Find radar beam-pointing vector, derived from the satellite altitude (roll, yaw, pitch angle).
3. Locate target center position, derived from the satellite position and line of sight.
4. Locate each target's polygon position, derived from the target center position and target aspect angle.

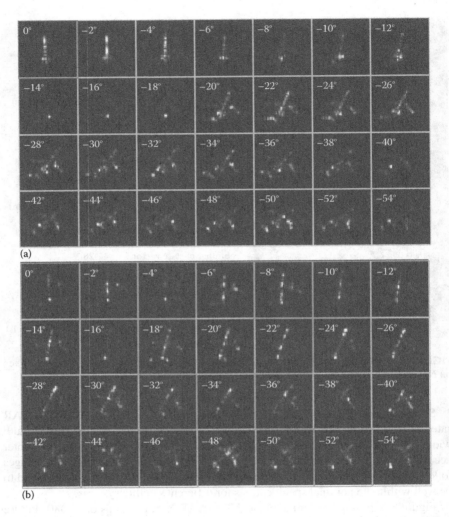

FIGURE 8.6 Simulated TerraSAR-X image clips of (a) MD80 and (b) B757 commercial aircraft with an incident angle of 30° and aspect angles of 0°–54°.

8.3 COMPUTATION COMPLEXITY

A SAR image is sensitive to a target's geometry, including orientation and aspect angles. For target recognition and identification, a more complete database for feature extraction is preferable to achieve better performance and reduce the false alarm rate. In SAR image simulation, suppose n samples (incident angles and aspect angles) are desired; then the computation complexity is $O(n^3)$. Table 8.2 lists CPU hours to complete one TerraSAR-X image simulation of an MD80 aircraft

Simulated results

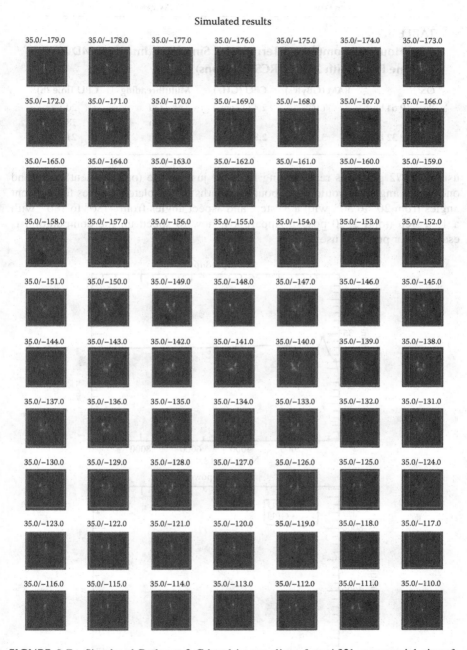

FIGURE 8.7 Simulated Radarsat-2 C-band image clips of an A321 commercial aircraft with an incident angle of 35° and aspect angles of −179° to 180°.

TABLE 8.2

CPU Hours to Complete a TerraSAR-X Simulated Image of MD80 for One Pose (with 25,672 RCS Polygons)

OS	RAM (GBytes)	CPU (GHz)	Multithreading	CPU Time (h)
Win-XP 64 bits	1.97	2.4 (4 cores)	Off	20.1
			On	18.7
Win-XP 32 bits	3.24	2.8 (2 cores)	Off	21.1

using 25,672 polygons representing RCS for just a pose (one incident angle and one aspect angle). It would take about 11 months to complete all poses for incident angles from 20° to 50°, with a 5° step, and aspect angles from −180° to 180°, with a 1° a step (total 2520 poses). Apparently, how to speed up the computation is essential for practical use.

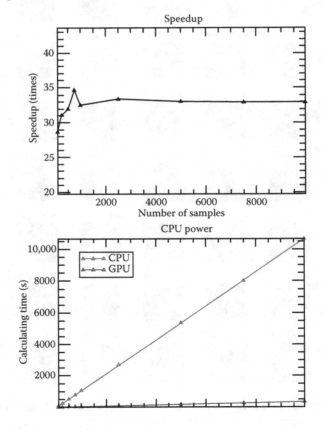

FIGURE 8.8 Speedup performance as a function of the number of polygons using a GPU algorithm.

As for the graphic processing unit (GPU), the work load is assumed to be highly parallel—many data to be processed by the same algorithm. Based on that assumption, each processing unit is designed to handle many threads. Processing units work as a group to maximize the throughput of all threads. Latency can be hidden by skipping stalled threads, if there are a few compared to the number of eligible threads. Grouping means shared control logics and cache. The GPU-based SAR simulation is divided into a grid of blocks. Each block consists of a number of *threads*, which are executed in a *multiprocessor* (also named stream multiprocessor or block). The Compute Unified Device Architecture (CUDA), developed by NVIDIA, is used for the proposed GPU-based implementation on a Linux operating system. The usage of the shared memory implemented in the proposed method is applied to the data-intensive computational tasks of the GPU-based computations. The data with high dependency are assigned to the same block of the shared memory with a finer granularity of parallel implementation. To make use of this highly efficient memory architecture, we devised local variables and parameters with these registers. The GPU-based experiments were performed on a low-cost 960 (240)-core NVIDIA Tesla personal supercomputer with one Intel Xeon 5504 quad-core CPU and four (one) NVIDIA GTX295 (240-core) GPUs. Figure 8.8 shows the speed-up performance using the GPU algorithm against a CPU one, as a function of the number of samples (polygons in the target's RCS) for one pose. As the number of samples increases, the CPU time grows very quickly. The required computation time is absorbed by GPU, as obviously seen from the chart. The average speedup ratio against the CPU is about 32. With a dual GPU configuration, a 65 times speedup was boosted. The speedup is of course keeping up with the progress of the advance of GPU power.

8.4 CONTRAST ENHANCEMENT BY NONQUADRATIC OPTIMIZATION

Good feature enhancement is essential for target identification and recognition of SAR images [5–8]. A novel algorithm proposed for Spotlight mode SAR [9,10] involves the formation of a cost function containing nonquadratic regularization constraints. Excellent results were reported. However, it is desirable to modify that approach to handle the Stripmap mode SAR data for their wide applications in earth observation [2,11].

In Stripmap mode, neither the radar antenna nor the target being mapped rotates, and the radar flight track is almost perpendicular to the look direction, with a small squint angle. To enhance the Stripmap mode data using the nonquadratic regularization method [11], one has to modify the projection operator kernel accordingly. In what follows, we outline the key steps to come up with a compact matrix equation for optimal estimation search, by mathematically reformulating the projection kernel and putting it into a form that is suitable for optimization. The performance was evaluated by measures of the target contrast and 3 dB beamwidth. Results were analyzed and compared with those using minimum variance (MV) [12] and multiple signal classification (MUSIC) [6] methods. Results demonstrate that the target's features are effectively enhanced and the dominant scattering centers are well separated

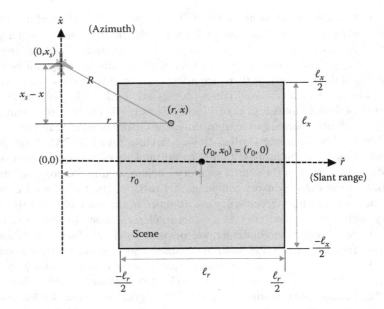

FIGURE 8.9 Stripmap mode received signal model in the azimuth–slant range plane.

using the proposed method. In addition, the image fuzziness is greatly reduced and the image fidelity is well preserved.

Figure 8.9 illustrates the SAR imaging geometry for the Stripmap mode, where (x_0, r_0) denote the scene center coordinates within the antenna footprint with slant range r_0, and $x_s(t_s)$ is the SAR position at time t_s along the azimuthal direction. Recall from Equation 1.54 that the transmitted signal is of the chirp form, with chirp rate a, carrier frequency ω_c, and a duration of T_p. Because radar measures time delay, it is more natural to illustrate the imaging in the slant plane with dimension $\ell_r \times \ell_x$, assuming for simplicity that the footprint formed by the antenna beam patterns is rectangular. The slant range from a target at (x, r) within the footprint to the radar at $(t_s, x_s = vt_s)$ is $R = R(x) = \sqrt{(x_s - x)^2 + r^2}$. Generally, $r^2 \gg (x_s - x)^2$, such that R can be approximated as

$$R \cong r + \frac{(x_s - x)^2}{2r} \tag{8.1}$$

From Equation 2.57, the return signal at the receiver from the complex scattered field $\mathbf{g}(x, r)$ is

$$s_{r0}(x - x_s, t - t_s) = \int_{-\frac{\ell_r}{2}}^{\frac{\ell_r}{2}} \int_{-\frac{\ell_x}{2}}^{\frac{\ell_x}{2}} \mathbf{g}(x, r) e^{-i\left\{\omega_c\left(t - \frac{2R}{c}\right) + \omega\left(t - \frac{2R}{c}\right)^2\right\}} \, dx \, dr \tag{8.2}$$

It is noted that the reference signal for the dechirping operation is the complex conjugate of the transmitted signal.

$$ref_{strip} = s_t^* = e^{-i[\omega_c t + at^2]}, \, t \in [-T_p/2, T_p/2] \tag{8.3}$$

where * indicates complex conjugate operation. Now by quadrature demodulation, the demodulated signal from the target $g(x, r)$ may be readily written as

$$s_r(x_s, t) = \int_{-\frac{\ell_r}{2}}^{\frac{\ell_r}{2}} \int_{-\frac{\ell_x}{2}}^{\frac{\ell_x}{2}} g(r, x) e^{-i\left[\omega_c \frac{2R}{c} - a\left(t - \frac{2R}{c}\right)^2\right]} dx \, dr \tag{8.4}$$

8.4.1 ALGORITHM FOR STRIPMAP MODE SAR BY NONQUADRATIC REGULARIZATION

The received signal s_r may be expressed as [9–11]

$$s_r = Tg + s_n \tag{8.5}$$

where T is the projection operation kernel with dimension $MN \times MN$, which plays a key role in contrast enhancement if the input signal, Equation 8.5, is of dimension $M \times N$. It has been shown that [9,10] nonquadratic regularization (NQR) is practically effective in minimizing the clutter while emphasizing the target features via

$$\hat{g} = \arg \min \left\{ \left\| s_r - Tg \right\|_2^2 + \lambda^2 \left\| g \right\|_p^p \right\} \tag{8.6}$$

where $\| \ \|_p$ denotes ℓ_p-norm ($p \leq 1$), λ^2 is a scalar parameter, and $\left\{ \left\| s_r - Tg \right\|_2^2 + \lambda^2 \left\| g \right\|_p^p \right\}$ is recognized as the cost or objective function.

To easily facilitate the numerical implementation, both s_r and g may be formed as long vectors, with T a matrix. Then, from Equations 8.4 and 8.5, we may write the projection operation kernel for the Stripmap mode as

$$T = \exp \left\{ -i \left[\omega_c \frac{2\left(r + \frac{(x-x_0)^2}{2r}\right)}{c} - a\left(t - \frac{2\left(r + \frac{(x-x_0)^2}{2r}\right)}{c}\right)^2 \right] \right\} \tag{8.7}$$

where $t, r, x - x_0$ are matrices to be described below by first defining the following notations to facilitate the formation of these matrices:

N: The number of discrete sampling points along the slant range direction
M: The number of discrete sampling points along the azimuth direction

Δt: The sampling interval along the slant range

ℓ_r: The size of the footprint along the slant range

ℓ_x: The size of the footprint along the azimuth direction

$\mathbf{1}_\ell$: The column vector of dimension $\ell \times 1$ and all elements equal to 1, in which $\ell = M$ or $\ell = MN$

With these notions, we can obtain explicit $\mathbf{t}$, $\mathbf{r}$, $\mathbf{x} - \mathbf{x}_0$ forms, after some mathematical derivations.

$$\mathbf{t} = [\mathbf{1}_M \otimes \mathbf{M}_1]_{MN \times MN} \tag{8.8}$$

$$\mathbf{v}_1 = \begin{bmatrix} 0 \\ 1 \\ \vdots \\ N-1 \end{bmatrix} \Delta t = \begin{bmatrix} 0 \\ \Delta t \\ \vdots \\ (N-1)\Delta t \end{bmatrix}_{N \times 1} \tag{8.9}$$

$$\mathbf{M}_1 = \mathbf{V}_1 \cdot \mathbf{1}_{MN}^T = \begin{bmatrix} 0 & \cdots & 0 \\ \Delta t & & \Delta t \\ \vdots & & \vdots \\ (N-1)\Delta t & \cdots & (N-1)\Delta t \end{bmatrix}_{N \times MN} \tag{8.10}$$

$$\mathbf{r} = \left[\mathbf{M}_2 \otimes \mathbf{V}_2^\dagger \right]_{MN \times MN} \tag{8.11}$$

$$\mathbf{V}_2 = \begin{bmatrix} \dfrac{-\ell_r}{2} + r_0 \\ \dfrac{-\ell_r}{2} + r_0 + \Delta\ell \\ \vdots \\ \dfrac{-\ell_r}{2} + r_0 + (N-1)\Delta\ell \end{bmatrix}_{N \times 1} , \quad \Delta\ell = \dfrac{\ell_r}{N} \tag{8.12}$$

$$\mathbf{M}_2 = \mathbf{1}_M^T \otimes \mathbf{1}_{MN} = \begin{bmatrix} \mathbf{1}_{MN} & \cdots & \mathbf{1}_{MN} \end{bmatrix}_{MN \times MN} \tag{8.13}$$

$$\mathbf{x} - \mathbf{x}_0 = [\mathbf{M}_3 \otimes \mathbf{M}_4]_{MX \times MN} \tag{8.14}$$

$$\mathbf{V}_3 = \begin{bmatrix} 0 \\ 1 \\ \vdots \\ M-1 \end{bmatrix} \frac{\ell_x}{M} - \frac{\ell_x}{2} = \begin{bmatrix} \dfrac{-\ell_x}{2} \\ \dfrac{\ell_x}{M} - \dfrac{\ell_x}{2} \\ \vdots \\ (M-1)\dfrac{\ell_x}{M} - \dfrac{\ell_x}{2} \end{bmatrix}_{M \times 1} \tag{8.15}$$

$$\mathbf{V}_4 = \begin{bmatrix} \mathbf{V}_3[\delta - 1 : (M-1)]_{\delta \times 1} \\ \mathbf{0}_{(M-\delta) \times 1} \end{bmatrix}_{M \times 1}, \delta = \left\lfloor \frac{M-1}{2} \right\rfloor + 1 \tag{8.16}$$

$$\mathbf{V}_5 = \begin{bmatrix} \text{Mirror}(\mathbf{V}_3[0 : \delta - 1])_{\delta \times 1} \\ \mathbf{0}_{(M-\delta) \times 1} \end{bmatrix}_{M \times 1} \tag{8.17}$$

$$\mathbf{M}_3 = \left[\text{Toep}\left(\mathbf{V}_5^T, \mathbf{V}_4^T \right) \right]_{M \times M} \tag{8.18}$$

$$\mathbf{M}_4 = \left[\mathbf{1}_{N \times 1} \cdot \mathbf{1}_{N \times 1}^T \right]_{N \times N} \tag{8.19}$$

In Equation 8.19, each of the bold-faced letters denotes a matrix, and $\otimes$ represents the Kronecker product. The operation $\mathbf{V}[m : n]$ takes element m to element n from vector $\mathbf{V}$, and Toep($\cdot$) converts the input into a Toeplitz matrix [13–15]. Note that in Equation 8.19, by Mirror[$a : b, c : d$]$_{p \times q}$ we mean taking elements a to b along the rows and elements c to d along the columns, so that the resulting matrix is of size $p \times q$.

Now that with the shorthand notations in Equations 8.9 through 8.19, the kernel $\mathbf{T}$ in the matrix is readily constructed, and thus the search for the optimum $\hat{\gamma}$ is in order. Readers should refer to [9–11] for details in this regard.

8.4.2 PERFORMANCE CHECK

To validate the effectiveness and efficiency of the procedure outlined above, real SAR images from a series of Radarsat SAR images acquired during a fine mode were tested with ground truth collected during the image acquisitions. Two figures of merit were used to evaluate the performance: target contrast and 3 dB beamwidth. Target contrast is defined as the ratio of the target amplitude and the surrounding background clutter amplitude within the region of interest, namely,

$$TC = 20\log_{10}\left\{ \frac{\max\limits_{(i,j)\in\text{Target}}\left|\hat{\gamma}_{i,j}\right|}{\dfrac{1}{N}\sum\limits_{(i,j)\in\text{Clutter}}\left|\hat{\gamma}_{i,j}\right|} \right\} \qquad (8.20)$$

Apparently, the higher the value of TC, the better is the feature enhancement. Another index to be used is a 3 dB beamwidth, $\vartheta_{3\text{dB}}$, as illustrated in Figure 8.10.

A 3 dB beamwidth is a focusing indicator of the target's scattering centers being extracted or enhanced. It is particularly important for the moderate-spatial-resolution SAR offered by most Stripmap mode systems. Hence, a good feature enhancement should simultaneously achieve large TC and narrow 3 dB beamwidth. To establish a benchmark reference, the popular enhancement algorithm's minimum variance (MV) [2] and multiple signal classification (MUSIC) [16] were tested for comparative evaluation.

For purposes of comparison, the MV and MUSIC methods are applied to enhance the SAR features. Suppose that the range profile processed from the range–Doppler focusing is of size $M \times N$; then, as given in Equation 8.21, the necessary correlation matrix $\mathbf{C}$ of the size $MN \times MN$ in MV can be derived from the column long vector with a size of $MN \times 1$. By the MV method, the enhanced column long vector γ is obtained as

$$\gamma_{k,l}^{MV} = \frac{1}{\mathbf{V}_{k,l}^{\mathbf{H}}\mathbf{C}^{-1}\mathbf{V}_{k,l}}, \quad k = 1,2,\ldots M; \ l = 1,2,\ldots N \qquad (8.21)$$

wherein H is the Hermitian operator; the $\mathbf{V}$ is the column long vector form of the matrix $W_{k,l}$:

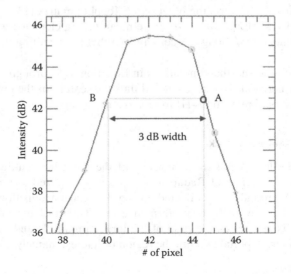

FIGURE 8.10 Definition of a 3 dB beamwidth for feature enhancement.

$$\mathbf{W}_{kl} = \frac{1}{\sqrt{M \times N}} \exp\left\{i2\pi\left(\frac{mk}{M} + \frac{nl}{N}\right)\right\}, \quad m \in [0, M-1]; \ n \in [0, N-1] \quad (8.22)$$

Finally, the destination or the required enhanced matrix is converted from the enhanced column long vector, γ, to matrix form.

By the same procedure, in MUSIC the enhanced vector γ is of the form

$$\gamma_{k,l}^{\text{MUSIC}} = \frac{\mathbf{V}_{k,l}^H \mathbf{V}_{k,l}}{\mathbf{V}_{k,l}^H \mathbf{e} \mathbf{e}^H \mathbf{V}_{k,l}} \quad (8.23)$$

where $\mathbf{e}$ is the matrix of the eigenvector of $\mathbf{C}$ with dimension $MN \times (MN - n_e)$; n_e is the dimension of the eigenvector in the output signal. Since there is no analytical method to obtain the n_e value, we determined it by trial and error.

Before proceeding to testing the real-world Stripmap SAR image, it is constructive to perform the enhancement using public MSTAR high-resolution data sets [17]. The resolution data in this experiment are 1 ft^2 at the X-band, and the radar beam direction was directed from the bottom to the top in Spotlight mode. It should be mentioned that since the enhancement is started from the signal data (before focused), instead of processing the whole data set to save computer time, the enhancement procedure is applied to the raw data of a region of interest containing the targets selected from the focused image. Figure 8.11 shows the enhanced images using MV, MUSIC, and the proposed method. Comparisons of the figures of merit, TC, and $\vartheta_{3\text{dB}}$ are given in Table 8.3. Compared to the original-magnitude image, the NQR method offers excellent image enhancement in suppressing the background noise while preserving the target features.

Now, we test on Stripmap SAR data from Radarsat-1 F3D mode SAR over a harbor on September 5, 2004, where each ship was labeled by s letter from A to R. The ships' positions were identified aided by the harbor vessel management system, and their photos were taken from two posts marked as ☆ (Figure 8.12) overflown with SAR imaging. For the purposes of this study, four targets (A, C, H, K) were selected for testing because they show visually different geometric features. The range profile data covering the interesting target was roughly extracted, and the slant range distance and time at each echo line were subsequently recorded. The size of each subimage containing the ships is 256 × 256 pixels.

Figure 8.13 displays the enhanced results using the presented method, along with the original image. The intensity histograms in the middle row of the figure clearly show strong separation of background clutter from the desired target. With the original image, it is almost impossible to tell a target from its background. The plots shown in the bottom row are to illustrate the clutter suppression: the clutter was greatly suppressed from −40 dB (right column) to −100 dB (left column). Distinctive scattering centers were also well identified, thus narrowing the beamwidth. Although not shown here, results were equally well produced from other ship targets, as given in Table 8.3 for TC and $\vartheta_{3\text{dB}}$.

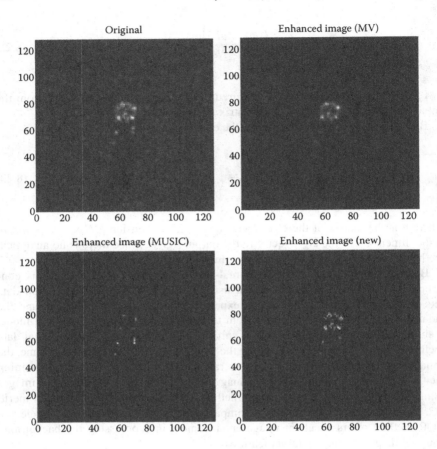

FIGURE 8.11 Comparison of the feature enhancement using MV, MUSIC, and the proposed method.

Table 8.3 summarizes a comparison of enhancement results using MV, MUSIC, and the NQR method, with target contrast (TC) and 3 dB beamwidth (ϑ_{3dB}) as figures of merit. In all cases under testing, it was found that the proposed method outperformed the other two popular methods in maximizing the target contrast and minimizing ϑ_{3dB}. Interesting to note is that MV gave results comparable to those of TC, but wider beamwidth when compared with the proposed method. On the other hand, MUSIC produced narrow beamwidth, but only low TC. Thus, each of the traditional methods was able to offer desirable outcomes on one of the measures, but neither was able to maximize both simultaneously. By reformulating the projection kernel and using it in an optimization equation form, an optimal estimate of the target's scattered or reflectivity field was obtained. As a result, the image fuzziness was greatly reduced and image fidelity was preserved. Thus, the target's features were adequately enhanced, and dominant scatterers were well separated.

TABLE 8.3

Performance Comparison of Different Enhancement Methods with Target Contrast and 3 dB Beamwidth as Figures of Merit

Target	Methods	TC (dB)		ϑ_{3dB} (pixels)	
		Original	Enhanced	Original	Enhanced
A	MV	23.3916	35.5828	1.86178	1.84425
	MUSIC		20.5336		0.883606
	NQR		45.9573		0.101635
H	MV	19.1605	32.2679	15.4995	1.45782
	MUSIC		17.8722		1.30253
	NQR		56.5543		0.0633373
K	MV	24.3508	37.2620	1.98624	1.89217
	MUSIC		26.3459		1.18142
	NQR		50.2857		0.0721550
C	MV	3.2898	30.2921	15.58981	1.50502
	MUSIC		11.9214		0.418796
	NQR		48.0110		0.0644417
MSTAR	MV	14.2773	28.8364	1.95905	1.33854
	MUSIC		24.8719		1.29002
	NQR		50.2519		0.44875

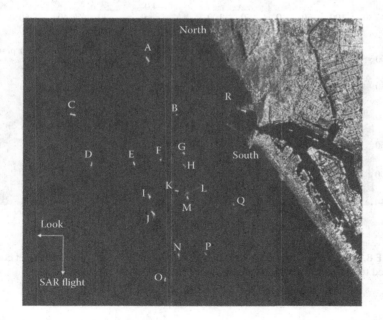

FIGURE 8.12 Radarsat-1 F3D mode SAR over a harbor on September 5, 2004. Each ship was labeled by a letter from A to R.

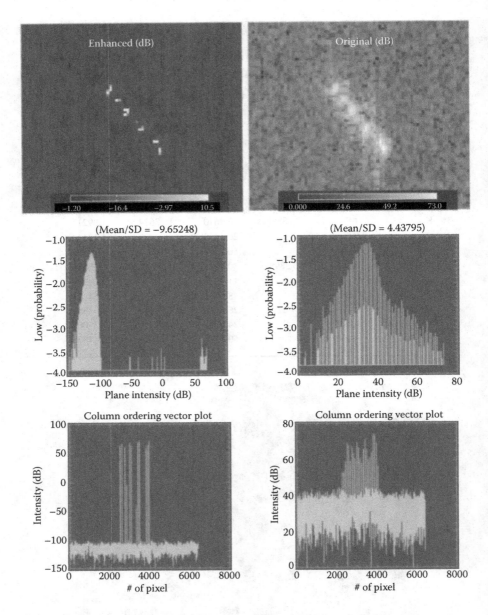

FIGURE 8.13 Original and enhanced images with intensity histograms and target contrasts displayed to illustrate the significant difference after the image was enhanced.

8.5 FEATURE VECTORS AND EXTRACTION

As an example, the feature vector we consider here contains two types: fractal geometry and scattering characteristics. In the fractal domain, the image is converted into a fractal image [18]. It has been explored that SAR signals may be treated as a spatial chaotic system because of the chaotic scattering phenomenon [19,20]. Applications of fractal geometry to the analysis SAR are studied in [21]. There are many techniques proposed to estimate the fractal dimension of an image. Among them, the wavelet approach proves both accurate and efficient. It stems from the fact that the fractal dimension of an N-dimensional random process can be characterized in terms of fractional Brownian motion (fBm) [18]. The power spectral density of fBm is written as

$$P(f) \propto f^{-(2H+D)} \tag{8.24}$$

where $0 < H < 1$ is the persistence of fBm, and D is the topological dimension (2 in image). The fractal dimension of this random process is given by $D = 3 - H$. As image texture, a SAR fractal image is extracted from SAR imagery data based on the local fractal dimension. Therefore, a wavelet transform can be applied to estimate the local fractal dimension of a SAR image. From Equation 8.25, the power spectrum of an image is therefore given by

$$P(u,v) = \upsilon\left(\sqrt{u^2 + v^2}\right)^{-2H-2} \tag{8.25}$$

where υ is a constant. Based on the multiresolution analysis, the discrete detailed signal of an image I at resolution level j can be written as [22]

$$
\begin{aligned}
D_j I &= \langle I(x,y), 2^{-j}\Psi_j(x - 2^{-j}n, y - 2^{-j}m)\rangle \\
&= (I(x,y) \otimes 2^{-j}\Psi_j(-x,-y))(2^{-j}n, 2^{-j}m)
\end{aligned}
\tag{8.26}
$$

where $\otimes$ denotes a convolution operator, $\Psi_j(x,y) = 2^{2j}\Psi(2^jx, 2^jy)$, and $\Psi(x, y)$ is a two-dimensional wavelet function. The discrete detailed signal can thus be obtained by filtering the signal with $2^{-j}\Psi_j(-x, -y)$ and sampling the output at a rate 2^{-j}. The power spectrum of the filtered image is given by [22]

$$P_j(u,v) = 2^{-2j} P(u,v)\left|\tilde{\Psi}_j(u,v)\right|^2 \tag{8.27}$$

where $\tilde{\Psi}_j(u,v) = \tilde{\Psi}(2^{-j}u, 2^{-j}v)$ and $\tilde{\Psi}(u,v)$ is the Fourier transform of $\Psi(u,v)$. After sampling, the power spectrum of the discrete detailed signal becomes

$$P_j^d(u,v) = 2^j \sum_k \sum_\ell P_j(u + 2^{-j}2k\pi, v + 2^{-j}2\ell\pi) \tag{8.28}$$

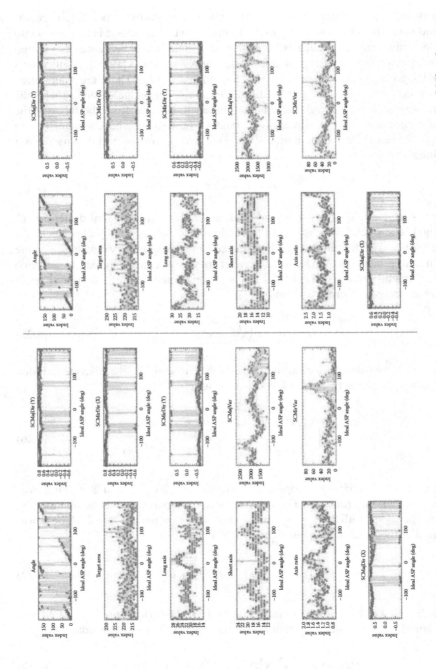

FIGURE 8.14 Selected features of simulated Radarsat-2 images of A321 (left) and B757-200 (right) for an incidence angle of 35° and aspect angles from −180° to 179° (1° step).

Let σ_j^2 be the energy of the discrete detailed signal:

$$\sigma_j^2 = \frac{2^{-j}}{(2\pi)^2} \iint P_j^d(u,v)\, du\, dv \tag{8.29}$$

By inserting Equations 8.27 and 8.28 into Equation 8.29 and changing variables in this integral, Equation 8.29 may be expressed as $\sigma_j^2 = 2^{-2H-2}\sigma_{j-1}^2$. Therefore, the fractal dimension of a SAR image can be obtained by computing the ratio of the energy of the detailed images:

$$D = \frac{1}{2}\log_2 \frac{\sigma_j^2}{\sigma_{j-1}^2} + 2 \tag{8.30}$$

A fractal image indeed represents information regarding spatial variation; hence, its dimension estimation can be realized by sliding a preset size window over the entire image. The selection of the window size is subject to reliability and homogeneity considerations, with the center pixel of the window replaced by the local estimate of the fractal dimension. Once the fractal image is generated, the features of angle, target area, long axis, short axis, and axis ratio are extracted from the target of interest. As for the scattering center (SC), the features of major direction (X), major direction (Y), minor direction (X), minor direction (Y), major variance, and minor variance are selected, in addition to the radar look angle and aspect angle. Figure 8.14 displays such a feature vector from simulated MD80 and B757-200 aircrafts from Radarsat-2. Finally, the recognition is done by a dynamic learning neural classifier [23–27]. This classifier is structured with a polynomial basis function model. A digital Kalman filtering scheme [28] is applied to train the network. The necessary time to complete training is basically not sensitive to the network size and is fast. Also, the network allows recursive training when new, and updated training data sets are available without revoking training from scratch. The classifier is structured with 13 input nodes to feed the target features, 350 hidden nodes in each of four hidden layers, and 4 output nodes representing four aircraft targets (A321, B757-200, B747-400, and MD80).

8.6   APPLICATIONS TO TARGET IDENTIFICATION

8.6.1   Test-Simulated SAR Images

Both simulated Radarsat-2 and TerraSAR-X images for four targets, as described above, are used to evaluate the classification and recognition performance. For this purpose, Radarsat-2 images with an incidence angle of 45° and TerraSAR-X images with an incidence angle of 30° were tested. Both are with s spatial resolution of 3 m in the Stripmap mode. The training data contains all 360 aspect angles (a 1° step); among them 180 samples were randomly taken to test. The fast convergence of neural network learning is observed. The confusion matrices for Radarsat-2 and

TABLE 8.4

Confusion Matrix of Classifying Four Targets from Simulated Radarsat-2 Images

		A321	B747-400	B757-200	MD80	Producer Accuracy
Target	A321	165	2	6	7	0.917
	B747-400	0	175	3	2	0.972
	B757-200	4	1	164	11	0.911
	MD80	4	2	11	163	0.906
User accuracy		0.954	0.972	0.891	0.891	

Overall accuracy, 0.926; kappa coefficient, 0.902.

TABLE 8.5

Confusion Matrix of Classifying Four Targets from Simulated TerraSAR-X Images

		A321	B747-400	B757-200	MD80	Producer Accuracy
Target	A321	166	0	5	9	0.922
	B747-400	0	179	0	1	0.994
	B757-200	2	0	168	10	0.933
	MD80	1	0	8	171	0.950
User accuracy		0.954	0.982	1.000	0.928	0.895

Overall accuracy, 0.950; kappa coefficient, 0.933.

TerraSAR-X in classifying four targets are given in Tables 8.4 and 8.5, respectively. The overall accuracy and kappa coefficient are very satisfactory. Higher confusion between B757-200 and MD80 was observed. It has to be noted that with the Radarsat-2 and TerraSAR-X systems, comparable results were obtained as long as the classification rate is concerned.

8.6.2 TEST ON REAL SAR IMAGES

Data acquisition on May 8, 2008, from TerraSAR-X over an airfield was processed into 4 look images with a spatial resolution of 3 m in the Stripmap mode. Feature enhancement by NQR was performed. Meanwhile, ground truth collections were conducted to identify the targets. Figure 8.15 displays such acquired images where the visually identified targets are marked and their associated image chips were later fed into a neural classifier that is trained by the simulated image database. Among all the targets, three types of them are contained in a simulated database: MD80, B757-200, and A321. From Table 8.6, these targets are well recognized, where the numeric value represents membership. A winner-takes-all approach was taken to

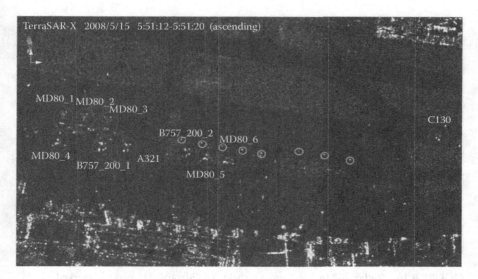

FIGURE 8.15 TerraSAR-X image over an airfield acquired on May 15, 2008. The identified targets are marked from the ground truth.

TABLE 8.6
Membership Matrix for Target Recognition on TerraSAR-X Image of Figure 8.15

			Output		
Input	**A321**	**B747-400**	**B757-200**	**MD80**	**Recognizable**
MD80_1	16.61%	7.26%	0%	76.12%	Yes
MD80_2	0%	9.63%	18.50%	71.85%	Yes
MD80_3	0%	3.63%	8.78%	87.58%	Yes
MD80_4	0%	3.64%	23.08%	73.27%	Yes
MD80_5	17.88%	12.28%	0%	69.82%	Yes
MD80_6	5.64%	0%	46.72%	47.63%	Yes
B757-200_1	12.05%	11.84%	76.09%	0%	Yes
B757-200_2	0%	19.41%	78.55%	2.03%	Yes
A321	63.28%	15.09%	0%	21.61%	Yes

determine the classified targets and whether they are recognizable. More sophisticated schemes, such as type II fuzzy, may be adopted in the future. It is realized that for certain poses, there exists confusion between MD80 and B757-200, as already demonstrated in the simulation test above.

As another example, a blind test was performed. Here blind means the targets to be recognized are not known to the tester, nor is the image acquisition time. The ground truth was collected by a third party and was only provided after the recognition operation was complete. This is very close to the real situation for recognition

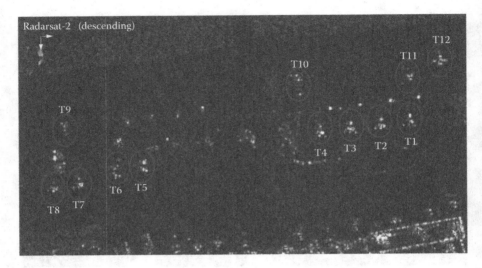

FIGURE 8.16 Radarsat-2 image over an airfield. The identified targets are marked.

operation. A total of 12 targets (T1–T12) were chosen for test, as indicated in Figure 8.16. Unlike in Figure 8.15, this image was acquired in descending mode, but was unknown at the time of test. Table 8.7 gives the test results, where the recognized targets and truth are listed. It is readily indicated that all the MD80 targets were successfully recognized, while four types of targets were wrongly recognized. The T12 target (ground truth was a C130) was completely not recognizable by the system. These are mainly attributed to the lack of a database. Consequently, enhancement and updating of the target database are clearly essential.

TABLE 8.7

Membership Matrix for Target Recognition on Raradsat-2 Image of Figure 8.16

	A321	B747-400	B757-200	MD80	Recognized	Truth
T1	19.25%	0%	22.88%	57.86%	MD80	MD80
T2	0.01%	0%	0%	99.98%	MD80	MD80
T3	14.99%	0%	11.59%	73.40%	MD80	MD80
T4	21.31%	10.81%	15.27	52.60%	MD80	MD80
T5	0%	3.09%	95.10%	1.79%	B757-200	E190
T6	96.82%	1.70%	0%	1.46%	A321	B737-800
T7	0.13%	0%	0.17%	99.69%	MD80	MD80
T8	0.10%	0.04%	0%	99.84%	MD80	MD80
T9	63.28%	15.09%	0%	21.61%	A321	FOKKER-100
T10	6.48%	2.26%	0%	91.24%	MD80	MD80
T11	43.61%	15.59%	40.59%	0%	A321	DASH-8

In this chapter, we presented a full-blown satellite SAR image simulation approach to target recognition. The simulation chain includes orbit state vector estimation, imaging scenario setting, target RCS computation, and a clutter model, all specified by a SAR system specification. As an application example, target recognition has been successfully demonstrated using the simulated image database for training sets. Extended tests on both simulated and real images were conducted to validate the proposed algorithm. To this end, it is suggested that a more powerful target recognition scheme should be explored for high-resolution SAR images. Extension to fully polarimetric SAR image simulation [29,30] seems highly desirable, as more such data are becoming available for much better target discrimination capability. In this aspect, further improvement on the computational efficiency by taking advantage of GPU power may be preferred and is subject to further analysis. Another direction toward the SAR simulation is to explore the image focusing based not on the point target model, but on volume scatter, which represents a more realistic physical mechanism, and thus the image will preserve more target information.

REFERENCES

1. Chen, C. H., ed., *Information Processing for Remote Sensing*, Singapore: World Scientific Publishing Co., 2000.
2. Chang, Y.-L., Chiang, C.-Y., and Chen, K.-S., SAR Image simulation with application to target recognition, *Progress in Electromagnetics Research*, 119: 35–37, 2011.
3. Chen, K.-S., and Tzeng, Y.-C., On SAR image processing: From focusing to target recognition, in *Signal and Image Processing for Remote Sensing*, ed. C. H. Chen, 2nd ed., 221–240, Boca, Raton, FL: World Scientific Publishing, 2012.
4. Cumming, I. G., and Wong, F. H., *Digital Processing of Synthetic Aperture Radar Data: Algorithms and Implementation*, Artech House, Boston, 2005.
5. Marcum, J. I., A statistical theory of target detection by pulsed radar, *IEEE Transactions on Information Theory*, 6: 59–267, 1960.
6. Rihaczek, A. W., and Hershkowitz, S. J., *Radar Resolution and Complex-Image Analysis*, Artech House, Norwood, MA, 1996.
7. Rihaczek, A. W., and Hershkowitz, S. J., *Theory and Practice of Radar Target Identification*, Artech House, Norwood, MA, 2000.
8. Trunk, G. V., and George, S. F., Detection of targets in non-Gaussian clutter, *IEEE Transactions on Aerospace and Electronic Systems*, 6: 620–628, 1970.
9. Çetin, M., and Karl, W. C., Feature-enhanced synthetic aperture radar image formation based on nonquadratic regularization, *IEEE Transactions on Image Processing*, 10(4): 623–631, 2001.
10. Çetin, M., Karl, W. C., and Willsky, A. S., A feature-preserving regularization method for complex-valued inverse problems with application to coherent imaging, *Optical Engineering*, 54(1): 017003-1–017003-11, 2006.
11. Chiang, C. Y., Chen, K. S., Wang, C. T., and Chou, N. S., Feature enhancement of Stripmap-mode SAR images based on an optimization scheme, *IEEE Geoscience and Remote Sensing Letters*, 6: 870–874, 2009.
12. DeGraaf, S., SAR imaging via modern 2-D spectral estimation methods, *IEEE Transactions on Image Processing*, 7(5): 729–761, 1998.
13. Bini, D., Toeplitz matrices, algorithms and applications, ECRIM News Online Edition, no. 22, July 1995.
14. Horn, R. A., and Johnson, C. R., *Topics in Matrix Analysis*, Cambridge, UK: Cambridge University Press, 1991.

15. Strang, G., *Introduction to Applied Mathematics*, Cambridge Press, Wellesley, 1986.
16. Schmidt, R. O., Multiple emitter location and signal parameter estimation, *IEEE Transactions on Antennas Propagation*, 34: 276–280, 1986.
17. Center for Imaging Science, MSTAR SAR database, available at http://cis.jhu.edu/data.sets/MSTAR/.
18. Mandelbrolt, A. B., and Van Ness, J. W., Fractional Brownian motion, fractional noises and applications, *IEEE Transactions on Medical Imaging*, 10: 422–437, 1968.
19. Leung, H., and Lo, T., A spatial temporal dynamical model for multipath scattering from the sea, *IEEE Transactions on Geoscience and Remote Sensing*, 33: 441–448, 1995.
20. Leung, H., Dubash, N., and Xie, N., Detection of small objects in clutter using a GA-RBF neural network, *IEEE Transactions on Aerospace and Electronic Systems*, 38: 98–118, 2002.
21. Tzeng, Y. C., and Chen, K. S., Change detection in synthetic aperture radar images using a spatially chaotic model, *Optical Engineering*, 46: 086202, 2007.
22. Mallat, S., *A Wavelet Tour of Signal Processing*, 2nd ed., Orlando, FL: Academic Press, 1999.
23. Chen, K. S., Tzeng, Y. C., Chen, C. F., and Kao, W. L., Land-cover classification of multispectral imagery using a dynamic learning neural network, *Photogrammetry Engineering and Remote Sensing*, 61: 403–408 1995.
24. Chen, K. S., Kao, W. L., and Tzeng, Y. C., Retrieval of surface parameters using dynamic learning neural network, *International Journal of Remote Sensing*, 16: 801–809, 1995.
25. Chen, K. S., Huang, W. P., Tsay, D. W., and Amar, F., Classification of multifrequency polarimetric SAR image using a dynamic learning neural network, *IEEE Transactions on Geoscience and Remote Sensing*, 34: 814–820, 1996.
26. Tzeng, Y., Chen, K. S., Kao, W. L., and Fung, A. K., A dynamic learning neural network for remote sensing applications, *IEEE Transactions on Geoscience and Remote Sensing*, 32: 1096–1102, 1994.
27. Tzeng, Y. C., and Chen, K. S., A fuzzy neural network to SAR image classification, *IEEE Transactions on Geoscience and Remote Sensing*, 36: 301–307, 1998.
28. Brown, R., and Hwang, P., *Introduction to Random Signal Analysis and Kalman Filtering*, Wiley, New York, 1983.
29. Lee, J. S., and Pottier, E., *Polarimetric Radar Imaging: From Basics to Applications*, CRC Press, Boca Raton, FL, 2009.
30. Margarit, G., Mallorqui, J. J., Rius, J. M., and Sanz-Marcos, J., On the usage of GRECOSAR, an orbital polarimetric SAR simulator of complex targets, to vessel classification studies, *IEEE Transactions on Geoscience and Remote Sensing*, 44: 3517–3526, 2006.

Index

Note: Page numbers ending in "f" refer to figures. Page numbers ending in "t" refer to tables.